Biomechanics

Other titles in the series

Biochemical genetics	R.A. Woods
Brain Biochemistry	H.S. Bachelard
Cell Differentiation	J.M. Ashworth
Cellular Development	D.R. Garrod
Functions of Biological Membranes	M. Davis
Immunochemistry	M.W. Steward
The Selectivity of Drugs	A. Albert

OUTLINE STUDIES IN BIOLOGY

General Editor: Professor J.M. Ashworth, University of Essex.

Editors' Foreword

The student of biological science in his final years as an undergraduate and his first years as a postgraduate is expected to gain some familiarity with current research at the frontiers of his discipline. New research work is published in a perplexing diversity of publications and is inevitably concerned with the minutiae of the subject. The sheer number of research journals and papers also causes confusion and difficulties of assimilation. Review articles usually presuppose a background knowledge of the field and are inevitably rather restricted in scope. There is thus the need for short but authoritative introductions to those areas of modern biological research which are either not dealt with in standard introductory textbooks or are not dealt with in sufficient detail to enable the student to go on from them to read scholarly reviews with profit. This series of books is designed to satisfy this need.

The authors have been asked to produce a brief outline of their subject assuming that their readers will have read and remembered much of a standard introductory textbook of biology. This outline then sets out to provide by building on this basis, the conceptual framework within which modern research work is progressing and aims to give the reader an indication of the problems, both conceptual and practical, which must be overcome if progress is to be maintained. We hope that students will go on to read the more detailed reviews and articles to which reference is made with a greater insight and interstanding of how they fit into the overall scheme of modern research effort and may thus be helped to choose where to make their own contribution to this effort.

These books are guidebooks, not textbooks. Modern research pays scant regard for the academic divisions into which biological teaching and introductory textbooks must, to a certain extent, be divided. We have thus concentrated in this series on providing guides to those areas which fall between , or which involve, several different academic disciplines. It is here that the gap between the textbook and the research paper is widest and where the need for guidance is greatest. In so doing we hope to have extended or supplemented but not supplanted main texts, and to have given students assistance in seeing how modern biological research is progressing, while at the same time providing a foundation for self help in the achievement of successful examination results.

Biomechanics

R. McNeill Alexander

Professor of Zoology
The University of Leeds

Chapman and Hall

London

First published in 1975
by Chapman and Hall Ltd.
11 New Fetter Lane, London EC4P 4EE
© 1975, R. McNeill Alexander
Typeset by E.W.C. Wilkins Ltd., London and Northampton, and
printed in Great Britain by William Clowes & Sons Ltd.,
London, Colchester and Beccles

ISBN 0 412 13080 7

Distributed in the U.S.A.
by Halsted Press, a Division
of John Wiley & Sons, Inc. New York

Library of Congress Catalog Card Number 75-2111

Contents

1	Introduction	*page*	7
	References		7
2	Biological materials		8
2.1	High polymers		8
2.2	Composite materials		11
2.3	Muscle		14
	References		16
3	Structure and materials		17
3.1	A dog on a force platform		17
3.2	Other organisms and other methods		19
3.3	Pennate muscles and tubular skeletons		22
	References		24
4	Animal locomotion		25
4.1	Power requirements		25
4.2	Running		28
4.3	Energy saved by elasticity		31
4.4	Flight		33
4.5	Swimming		36
	References		37
5	Human mechanics		39
5.1	Lubrication of joints		39
5.2	Blood		40
5.3	Surface tension in the lungs		43
	References		45
6	Plant mechanics		46
6.1	How sap rises		46

6.2	How stomata open	49
6.3	Falling seeds	50
	References	52
7	Cell mechanics	53
7.1	Cilia and flagella	53
7.2	Cell division	56
7.3	Myonemes	57
7.4	Cell membranes	58
	References	59
8	Conclusion	60
	References	60
	Index	61

1 Introduction

Biomechanics is the application of mechanics to the study of organisms. It crosses traditional subject boundaries and this naturally gives rise to some difficulties. Nearly all practitioners of biomechanics are biologists who have taught themselves some physics and engineering or engineers who have taught themselves biology. The biologists are apt to make quite elementary mistakes through misunderstanding some physical theory. The engineers are apt to overlook the complexity of biological structure. There is a lot to be gained by collaboration between biologists and engineers, but consultation is no substitute for learning. The biologist must make sure that his knowledge of mechanics is sound, even if it is elementary. The engineer must familiarize himself with the anatomy and physiology of the organisms he is studying. This book should show that the combination of biology and mechanics need not be forbiddingly difficult.

The book is designed to illustrate the range of biomechanics, to show where the subject stands at present and to point to some unsolved problems. Creative science is unpredictable so I cannot say where the next major advance will be made but I can report some of the questions which are being asked.

Many readers will be familiar with the S.I. system of units. A few who learnt their physics long ago may not, and this paragraph is for them. The basic units of length, mass and time are the metre, kilogramme and second. The unit of force is the Newton: a force of 1 N gives a mass of 1 kg an acceleration of 1 m s^{-2}. Since the acceleration due to gravity is 9.8 m s^{-2} a Newton is $1/9.8 = 0.102$ kg wt. Energy is measured in Joules: 1 J is the work done when a force of 1 N moves 1 m along its line of action. Power is measured in Watts, which are Joules per second. Multiples and fractions of the units are indicated by standard prefixes. Thus

$$1 \text{ GN} = 10^9 \text{ N} \qquad 1 \text{ MJ} = 10^6 \text{ J}$$

$$1 \text{ km} = 10^3 \text{ m} \qquad 1 \text{ mW} = 10^{-3} \text{ W}$$

$$1 \, \mu\text{s} = 10^{-6} \text{ s} \qquad 1 \text{ ng} = 10^{-9} \text{ g}$$

Readers may find the following general books useful, in addition to the books and papers listed at the ends of the chapters which follow.

References

[1] Alexander, R.McN. (1968), *Animal Mechanics*, Sidgwick and Jackson, London.
[2] Holwill, M.E.J. and Silvester, N.R. (1973), *Introduction to Biological Physics*. Wiley, London.
[3] Walshaw, A.C. (1970), *S.I. Units and Worked Examples*, Longman, London.
[4] Jarman, M. (1970), *Examples in Quantitative Zoology*, Arnold, London.
[5] White, D.C.S. (1974), *Biological Physics*, Chapman and Hall, London.

2 Biological materials

This chapter is about some of the materials animals and plants are made of. It deals mainly with their strength and elasticity, and it will be evident that our understanding of these properties is still very incomplete [15]. There is also a section about the contractile properties of muscle.

2.1. High polymers

Many of the structural materials in animals and plants are proteins or polysaccharides. Proteins (such as collagen, which forms tendons and ligaments) consist of large numbers of amino-acid residues joined together in long chains. Polysaccharides (such as the cellulose of plant cell walls) consist of long chains of sugar units. Materials like these, which have molecules built up from large numbers of more or less similar units, are known as high polymers. They have distinctive mechanical properties which depend more on the arrangement of the molecules than on the chemical nature of the units from which they are built. This is convenient, for rubbers and plastics are also high polymers and their mechanical properties have been studied intensively because of their importance in industry [14]. The theoretical understanding of high polymers which has been developed by rubber and plastics engineers can be applied by biologists to proteins and polysaccharides.

Fig. 2.1. represents small samples of polymers, with the long, flexible molecules illustrated as sinuous lines. It shows three distinctive types of polymer, and the effect of

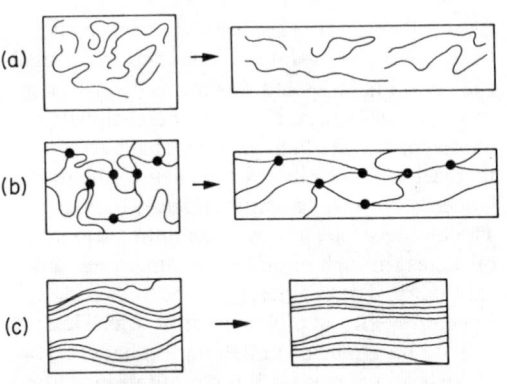

Fig. 2.1 A diagram showing the arrangement of the molecules in three types of high polymer, and the effects of stretching. (a) An amorphous polymer which is not cross-linked; (b) An amorphous cross-linked polymer and (c) A fibre.

stretching each. The molecules of the amorphous, non-cross-linked polymer (a) are separate from each other. They can slide past each other so the material can be stretched or otherwise distorted indefinitely. There is no tendancy for the material to return to its original shape; it is not elastic, but behaves as a liquid or a plastic solid.

Though they have no elasticity (apart from transient elastic properties due to the molecules getting tangled) amorphous, non-cross-linked materials often have high viscosity. The difference between elasticity and viscosity is a fundamental one. Both elastic and viscous materials resist deformation but in elastic ones (such as rubber) the resisting force depends on the amount of deformation while in viscous ones (such as treacle) it depends on the rate of deformation. Energy used deforming an elastic material is stored in the material and can be recovered in an elastic recoil. Energy used deforming a viscous material is lost as heat. The properties of elasticity and viscosity are not exclusive: indeed, all elastic materials are to some extent viscous.

Latex is an amorphous, non-cross-linked polymer and is a viscous liquid. It is converted into rubber (which is of course an elastic solid) by the process of vulcanization. This puts bridges of sulphur between the molecules, linking them together in a three-dimensional network. Rubber is a typical amorphous cross-linked polymer with the structure shown in Fig. 2.1b. It is hard and brittle like glass at very low temperatures, but at normal temperatures it is highly extensible. A rubber band can be stretched to three times its initial length, and will spring back when released.

The mechanism of the elasticity of rubbery polymers is quite different from that of other materials such as (for instance) steel. The different mechanism explains the remarkable extensibility. The long molecules are in Brownian motion, writhing all the time. At a given instant a particular molecule may be rolled up into a tight ball or stretched out straight, but it is a much more likely to be somewhere between these extremes. The length of the molecule keeps changing as it coils and uncoils at random but has a particular most probable value. When the material is stretched the molecules cannot slide past each other because of the cross links, but tend to be straightened out in the direction of stretching. When it is released random coiling and uncoiling tends to restore the molecules to their most probable length, and the material to its original dimensions.

If you open a scallop (*Pecten*) shell you will find inside, at the hinge, a block of material called abductin (Fig. 2.2). So long as it is wet,

abductin is extraordinarily like rubber. If you cut it out and drop it on the floor you will find it bounces well. It serves as a compression spring, to open the shell. There is a muscle which closes the shell and holds it closed but when the muscle relaxes the abductin makes the shell spring open. Abductin has properties very like rubber but it is an amorphous cross-linked protein.

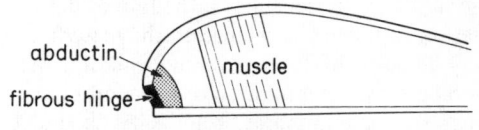

Fig. 2.2 A diagrammatic transverse section of a scallop (*Pecten*) showing the abductin and the muscle which closes the shell.

Resilin is another protein with similar properties [1]. It is amorphous and cross-linked, but the cross-links are chemically different from those of abductin. It is found in the thoraxes of insects and plays an important energy-saving role in the mechanism of flight (p. 37). It forms the elastic element in the catapult mechanism which enables fleas to jump high [3]. It is also present in the proboscis of butterflies and moths. This is a spiral tube which is uncoiled for use by applying pressure to the fluid within it. It is coiled up neatly again after use by elastic recoil of the resilin in its wall.

Dragonflies have an apodeme which is almost pure resilin. It is less than a millimetre long even in quite large dragonflies, but its rod-like shape made it reasonably convenient for investigations of mechanical properties. Torkel Weis-Fogh tied fine nylon threads to its ends and stretched it in a chamber of water under a microscope [1]. It could be stretched to three times its initial length. It could be left stretched for days but would still snap back to its original length when it was released. Investigations of the effect of temperature on its elastic properties provided evidence that the mechanism of its elasticity is the same as for rubber.

In one property resilin is superior to rubbers. This is the property of resilience. A ball which is dropped and bounces never rises quite to the height from which it fell. Part of the reason for this is that rubber and similar materials are to some extent viscous. Some of the energy used deforming them is needed to overcome viscosity and is lost as heat, so it cannot be recovered in the elastic recoil. The percentage recovered is the resilience. It depends on the rate of deformation and recoil, as well as on the material: it may be quite different when deformation and recoil are over in a fraction of a second than when they take minutes or hours. At the most favourable rates the resilience of rubbers is typically about 91%. In experiments in which locust resilin was deformed and released 50 times per second it was found to have a resilience of 97%. It will be apparent when the function of resilin in flight is explained (p. 36) that high resilience in rapid cycles of deformation and release is a valuable property.

Elastin is another elastic protein. The ligamentum nuchae which runs along the necks of ungulates above the vertebrae consists of elastin. It helps to support the head but is extensible enough to allow the animal to lower its head to graze. There is also elastin in the walls of arteries, especially near the heart. The arteries are distended by each heart beat, storing blood which is driven out of them again by elastic recoil in the interval between heart beats. Thus blood is kept flowing through the smaller blood vessels throughout the heart beat cycle.

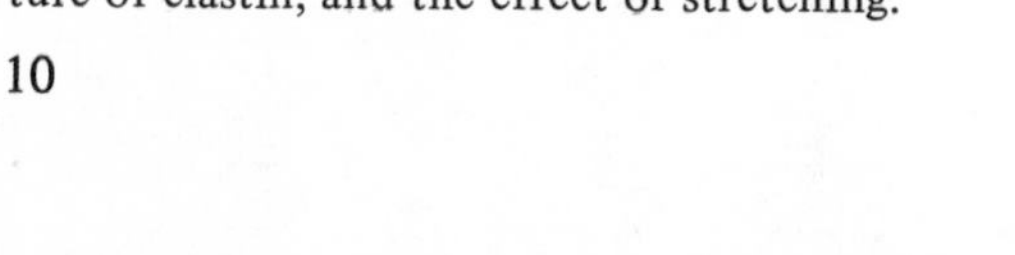

(a)   (b)

Fig. 2.3 A diagram showing the probable structure of elastin, and the effect of stretching.

It used to be generally believed that elastin is a straightforward amorphous cross-linked polymer like abductin, resilin and rubber. Recent research suggests a different structure and an entirely different mechanism of elasticity [16]. It seems that elastin may consist of globules of protein which are linked together but have water between them (Fig. 2.3). The globules are initially spherical but stretching draws them out into the shape of a Rugby (or American) football. This increases their surface area and so is opposed by surface tension. When stretched elastin is released, surface tension makes the globules become spherical again and so drives the elastic recoil. This hypothesis of the mechanism of elasticity was devised to explain the results of experiments on the effects of temperature, which are inconsistent with rubber-like elasticity, but there are other possible explanations [11].

The relationships between force and deformation for elastic materials are expressed by moduli of elasticity. There are various moduli relating to different types of deformation. The most familiar is Young's modulus, which is concerned with stretching or compression in one direction. It is tensile stress (i.e. force per unit cross-sectional area) divided by tensile strain (i.e. fractional change in length). Its value, for a cross-linked polymer, depends on the spacing of the cross-links: the shorter the length of molecule between one cross-link and the next, the higher the modulus (i.e. the harder the material is to stretch). Young's modulus is of the order of 1 MN m^{-2} for soft rubber, abductin and resilin. It happens to be about the same for elastin, though the mechnism of the elasticity is different.

There is another group of high polymers which have much higher Young's moduli, of the order of 1 or 10 GN m^{-2}. They cannot be stretched nearly as much as rubbers: they generally break when extended by 20% or even less. They are the fibres (the so-called fibres of elastin are not fibres in this technical sense). They are relatively inextensible because their molecules do not coil at random. X-ray diffraction studies show that they have crystalline regions where the moleculse are lined up parallel to each other and amorphous regions where they are not. Electron microscopy shows that most (but not all) fibres have fine curving fibrils running along their length. Fig. 2.1c shows one of several suggestions which have been made as to how the molecules are arranged in typical fibres. Only a little stretching is possible because the crystalline fibrils are soon straightened out. Even this stretching is resisted by elastic forces because it distorts the amorphous material between the fibrils.

Silk and collagen are protein fibres. Silk is formed as an amorphous, non-cross-linked polymer within the silkworm. It is liquid until it is drawn out through a nozzle. As the molecules pass through the nozzle they tend to line up along the direction of flow. Once lined up, they tend to attach to each other in crystalline fashion. The liquid is changed to a fibre simply by passing through the nozzle.

Collagen is not quite a typical fibre, because its fibrils have regular kinks [8]. The kinks straighten as it is stretched. Its fibrous structure is destroyed by heat: a tendon heated over 65°C shrinks to about a third of its initial length and becomes an amorphous, cross-linked polymer with rubber-like properties.

Cellulose and chitin are polysaccharide fibres. Cellulose is a polymer of glucose, and the principal constituent of plant cell walls. Chitin is similar, but has one of the hydroxyl groups in each glucose unit replaced by $-NHCOCH_3$. It is a major constituent of insect cuticle.

2.2 Composite materials

Wood and insect cuticle both consist of fibres embedded in another material. The grain of wood is the direction of the tracheids, which are long slender cells with very thick walls (Fig.

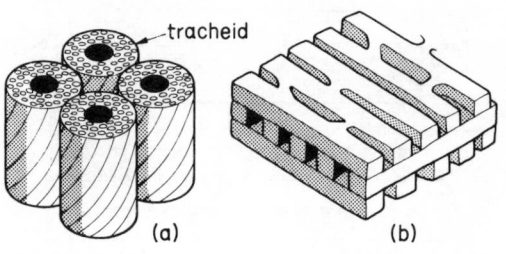

Fig. 2.4 Diagrams showing the structure of (a) wood and (b) beetle cuticle. Only the main (S_2) layer of fibres is represented in each tracheid in (a).

2.4a). The cellulose fibres in these walls run along the cells in slightly helical paths [13]. About 70% of the dry weight of typical softwood is cellulose, or related polymers known collectively as hemicellulose. Most of the rest is lignin, a branched amorphous polymer which fills the spaces between the cellulose fibres, and between one tracheid and the next. Insect cuticle consists of layers of parallel chitin fibres (Fig. 2.4b; Ref. [12]). Successive layers have their fibres running in different directions so the cuticle is more like plywood than solid wood. The spaces between the fibres are filled by protein which is cross-linked (in hard cuticle) by the process of quinone tanning.

Thus wood and insect cuticle are composite materials, like a number of artificial materials which have been invented in recent years [9]. One of the most useful of these materials is fibreglass, which consists of fine glass fibres embedded in a plastic resin. Glass is brittle and the resin is not particularly strong but fibreglass made from them is strong and tough enough to be used for making boats and fishing rods. Why can composite materials be superior to their constituents? It is obvious enough that a fabric of cellulose or chitin fibres will be stiffer if it is impregnated with some sort of glue, but might not a solid block of cellulose or chitin be even stiffer and stronger? The analogy of fibreglass suggests it might not be as strong.

11

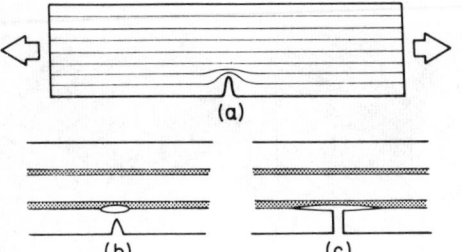

Fig. 2.5 These diagrams are explained in the text.

Breaking a material involves separating the two layers of atoms on either side of the break. Physical theory enables us to calculate the stresses required for this. It gives values of the order of 10 GN m^{-2} for glass, cellulose and steel, but the materials in bulk are nothing like as strong as this. The strength of ordinary glass is only 0.2 GN m^{-2}, though very fine, freshly drawn glass fibres can approach the theoretical strength quite closely. The strength of fibreglass can be up to about 1 GN m^{-2}.

Materials tend to break at stresses far below their theoretical strength because of the phenomenon of stress concentration which is illustrated in Fig. 2.5a. This diagram represents a bar with a notch in it under tension. The thin lines show the direction of the tension and where they are evenly spaced the stress is uniform right across the bar. They are diverted round the notch. They are close together, and the stress is high, at the tip of the notch. The stress here may be very many times as high as in the bulk of the material, depending on the sharpness of the notch, and fracture may occur here while the stress in the bulk of the material is still only a tiny fraction of the theoretical breaking stress. If this happens the crack formed at the tip of the notch will tend to spread right across the bar; a crack is in effect a very sharp notch, so high stresses will occur at its tip. A bar need not of course be deliberately notched, but surface flaws are inevitable in any material

subject to ordinary wear and tear so stress concentrations are bound to occur. Fine, freshly drawn glass fibres can have very perfect surfaces and can approach the theoretical strength but is is not practicable to make ordinary structures as perfect as this.

Composite structure can help to stop cracks spreading. A bar in tension has tensile stresses along it, but where the forces are diverted round a crack (Fig. 2.5a) the tensile stress has a component acting across the bar. Suppose the bar consists of strong fibres held together by weaker glue. Then as a crack approaches the boundary between two fibres this transverse component of stress may be enough to break the glue and pull the fibres apart (Fig. 2.5b). The crack may reach the boundary but its path will be interrupted there and it may not cross into the next fibre (Fig. 2.5c). This is a situation where the weakness of glue can, paradoxically, give strength to a material.

Many animal skeletons contain crystals of inorganic salts. Generally the crystals are relatively small and separated by protein, as in bone and in mollusc shells. The ossicles of echinoderms are exceptional [7]. Each ossicle is a three-dimensional network of bars of calcium carbonate. There are no boundaries between crystals: it is as though each ossicle were carved from a single crystal. This strange structure may help to prevent cracking. A solid crystal would be liable to crack right across but an ossicle is probably less likely to do so: a crack may cross one bar but is unlikely to go further.

Bone is a composite of collagen fibres and inorganic crystals [4], [15]. The crystals are tiny, about 20 nm long, and approximate in composition to the formula $3Ca_3(PO_4)_2 . Ca(OH)_2$. They seem to be firmly attached to the fibres, which are arranged in various ways. Commonly the bone is laid down around blood vessels, in concentric cylindrical layers (Fig. 2.6). The layers of bone around a single blood vessel are known as an osteone. Over substantial areas in each

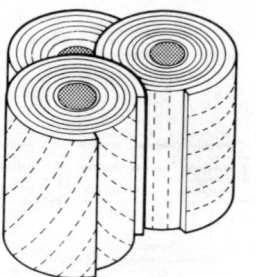

Fig. 2.6 A diagram illustrating the structure of osteone bone.

layer all the collagen fibres are parallel, but the direction of the fibres changes from layer to layer. Longitudinal fibres may alternate with circular ones in successive layers, or left-handed helices with right-handed ones. The layers are not necessarily equal in thickness so an osteone may consist predominantly of longitudinal or of circular fibres. One would expect osteones consisting mainly of longitudinal fibres to be strongest in tension and ones consisting mainly of circular ones to be strongest in compression. Tests on single osteones dissected out of bones have shown that this is indeed the case [2]. It seems likely that osteones with predominantly longitudinal fibres may tend to develop in parts of bones which are stressed mainly in tension, and ones with predominatly circular fibres in parts stressed mainly in compression. So far as I know, no one has looked to find out whether this is the case.

The inorganic crystals make bone considerably stiffer than collagen alone would be (Young's modulus is about 1 GN m^{-2} for collagen and $10 - 20 \text{ GN m}^{-2}$ for bone). They also contribute to the strength of bone. If they did not bone consisting of equal volumes of collagen and crystals would be only half as strong as pure collagen. In fact, bone seems generally to be rather stronger than collagen. It seems odd that such tiny crystals can strengthen bone, but a similar effect can be seen in the filled rubbers which are used for making tyres. These consist

of rubber mixed with fine soot. The soot not only stiffens the rubber, but makes it very much stronger.

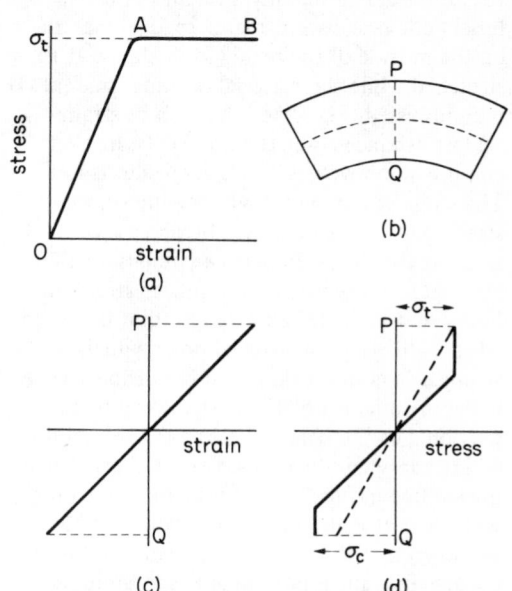

Fig. 2.7 (a) An idealized graph of stress against strain for a tough material. (b) A diagram of a bent bar. (c), (d) Graphs of strain and of stress, respectively, against distance along the line PQ in (b). σ_t, σ_c, tensile and compressive yield stresses. Further explanation is given in the text.

Fig. 2.7a shows how bone and other tough materials behave under stress. Stretching from O to A is elastic and reversible: if the material is stretched to A and released it will spring back to O. If it is stretched beyond A irreversible yielding occurs and though it will shorten when released it never quite returns to A. It breaks at B. Brittle materials such as glass yield very little but tough ones such as wet bone and ductile metals yield a good deal. The capacity to yield makes them tough. If dangerously high stresses arise in a stress concentration yielding occurs and spreads the stresses out more evenly. This

13

makes it less likely that a crack will start.

Yielding is particularly effective in reducing the danger of breakage when bone is being bent [5]. Consider an initially straight bar of material bent as shown in Fig. 2.7b. The material on the outside of the bend (for instance at P) is stretched while the material on the inside (at Q) is compressed. Fig. 2.7c is a graph of strain against distance along the line PQ (note that compression involves negative tensile strain). The stress at each point will depend on the strain, and the material will break as soon as the stress at P exceeds the tensile strength or the stress at Q exceeds the compressive strength. Suppose it is the tensile strength that is exceeded (as is likely in the case of bone, which is stronger in compression than in tension). Then if the material is brittle the stress will be proportional to the strain all along PQ and failure will occur when the stresses are as shown by the dotted line in Fig. 2.7d. If it is tough yielding will occur at P and the stresses will come to be as shown by the continuous line before fracture occurs. This apparently happens in bone. It makes cylindrical rods of bone about twice as strong in bending as they would be if no yielding occurred.

2.3 Muscle

Muscle is very different from the materials so far considered. They are passive, but it can use chemical energy to perform mechanical work [17]. There are various types of muscle which differ in structure and properties, but only vertebrate striated muscle will be described here. The muscles which move the skeletons of vertebrates are of this type but the muscles which move the viscera are not. Some crustacean and insect muscles are very similar.

Vertebrate striated muscle consists largely of filaments of two proteins. There are thick filaments of myosin and thin filaments of actin, and all run lengthwise along the muscle fibres (Fig. 2.8). They are arranged in a regular,

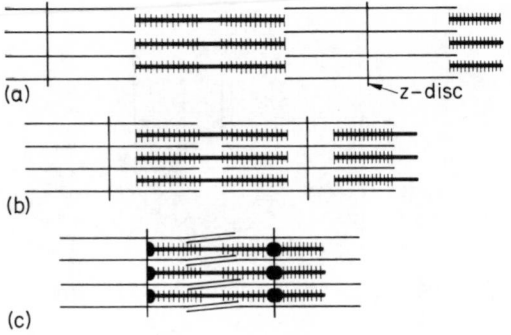

(a)

(b)

(c)

z–disc

Fig. 2.8 Drawing showing the arrangement of the filaments in a vertebrate striated muscle fibre: (a) when the fibre is greatly extended, (b) at an intermediate length and (c) when it is greatly contracted.

repeating pattern which is responsible for the stripes, running across the muscle fibres, which can be seen under the microscope. The filaments do not change their length as the muscle extends and contracts, but slide past each other as shown in the diagrams. Projections from the thick filaments can attach to the thin ones and exert forces on them: they are responsible for the force exerted by the muscle fibre, when it is active. If the fibre is stretched beyond the point represented in Fig. 2.8a none of the projections can attach to thin filaments and no force can be exerted. At the point represented in Fig. 2.8b all the projections can attach to thin filaments and the force which can be exerted is maximal. Shortening beyond this point brings the tips of the thin filaments beyond the mid-points of the thick filaments. Since all the projections pull towards the mid-point, this means that the forces on the tips of the thin filaments act in the wrong direction: a given thick filament pulls on one part of a thin filament and pushes on another, and the net force is reduced. Further shortening squashes the tips of the thick filaments against the z-discs which join the thin filaments together (Fig. 2.8c) and eventually no force can be exerted. The greatest length at which the fibre can exert a force (apart from small elastic forces due to stretching of the membrane which covers it) is almost three times the minimum length at which force can be exerted. The forces become small as the extremes are approached, and most striated muscles are arranged in the body so that the range of lengths occurring in normal movements is much smaller than this. Other types of muscle are different in structure and some can work over larger ranges of length.

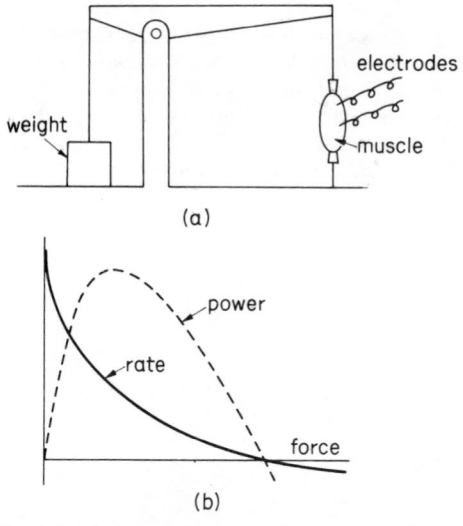

(a)

(b)

Fig. 2.9 (a) An experiment with living muscle which is described in the text and (b) data obtained by this experiment.

The information which has just been given is an interpretation of data obtained in an experiment with fibres from frog leg muscles [10]. A muscle fibre was dissected from a freshly killed frog and stretched to various lengths. It was stimulated electrically at each length and the force it exerted was measured. The width of the striations (stripes) on the fibres was observed through a microscope at the same time and the amount of overlap between the thick and thin filaments was inferred.

Another experiment with muscle is illustrated in Fig. 2.9a. This experiment can be performed either with single fibres or with whole muscles. The muscle is attached to one end of a lever which has a weight hanging from the other. It is stimulated by mild electric shocks and lifts the weight, provided the weight is not too heavy. A device which is not illustrated makes a record of movements of the lever which shows (as might be expected) that small weights are lifted faster than large ones. Fig. 2.9b includes a graph of the rate at which the muscle shortens against the force it has to exert. The force which just prevents the muscle from shortening is known as the maximum isometric force. Greater forces stretch the muscle: rates of stretching appear on the graph as negative rates of shortening.

The athletic performance of animals depends largely on the mechanical power their muscles can produce. The power output of a muscle is the rate of shortening multiplied by the force being exerted. Fig. 2.9b shows that it is greatest when the force is about 0·3 of the maximum isometric force: this has been found to be true of a wide range of muscles.

The stress exerted by a muscle fibre, like the stress in a passive material, is the force per unit cross-sectional area. The maximum isometric stress is the stress corresponding to the maximum isometric force. It seems generally to lie between 200 and 350 kN m^{-2} for vertebrate striated muscles at their optimum lengths: lower values obtained in experiments with some muscles may be due to inadequate stimulation, or deterioration of the muscle. The stress which allows maximum power output is 0·3 of the maximum isometric stress, or about 60–100 kN m^{-2}. Some invertebrate muscles can exert much larger stresses. The maximum rates at which muscles can shorten very widely [6].

References

[1] Andersen, S.O. and Weis-Fogh, T. (1964), *Adv. Insect Physiol.*, **2**, 1—65.

[2] Ascenzi, A. and Bonucci, E. (1968), *Anat. Rec.*, **161**, 377—92.

[3] Bennett-Clark, H.C. and Lucey, E.C.A. (1967), *J. exp. Biol.*, **47**, 59—76.

[4] Bourne, G.H. (ed.) (1972), *The Biochemistry and Physiology of Bone*, edition 2. Academic Press, New York.

[5] Burstein, A.H., Currey, J.D., Frankel, V.H. and Reilly, D.T. (1972), *J. Biomech.*, **5**, 35—45.

[6] Close, R. (1965), *J. Physiol.* Lond., **180**, 542—59.

[7] Currey, J.D. (1970), *Animal Skeletons*, Arnold, London.

[8] Diamant, J., Keller, A., Baer, E., Litt, M. and Arridge, R.G.C. (1972), *Proc. R. Soc. Lond.* B, **180**, 293—315.

[9] Gordon, J.E. (1968), *The New Science of Strong Materials, or Why you don't fall through the floor.* Penguin Books, Harmondsworth, Middlesex.

[10] Gordon, A.M., Huxley, A.F. and Julian, F.J. (1966), *J. Physiol. Lond.*, **184**, 170—92.

[11] Gray, W.R., Sandberg, L.B. and Foster, J.A. (1973) *Nature* Lond., **246**, 461—6.

[12] Hepburn, H.R. and Ball, A. (1973), *J. Materials Sci.*, 8, 618—23.

[13] Mark, R.E. (1967), *Cell Wall Mechanics of Tracheids*, Yale University Press, New Haven.

[14] Ritchie, P.D. (ed.) (1965), *Physics of Plastics*, Iliffe, London.

[15] Wainwright, S.A., Biggs, W.S., Currey, J.D. and Gosline, J.M. (in press) *Mechanical Design in Organisms*, Arnold, London.

[16] Weis-Fogh, T. and Andersen, S.O. (1970), *Nature, Lond.*, **227**, 718—21.

[17] Wilkie, D.R. (1970), *Muscle*, Arnold, London.

3 Structure and materials

The properties of biological materials are interesting in their own right but they are even more interesting when they can be related to the demands made on them in the bodies of animals and plants. This chapter describes how properties of materials considered in Chapter 2 are exploited in life.

3.1 A dog on a force platform

Each muscle in an animal body must be large enough for the most strenuous activity required of it. Each bone and tendon must be strong enough to withstand the forces exerted by the muscles. An investigation of jumping by dogs will serve as a convenient example of the kind of analysis that is possible [4].

The dog used in the experiments had been trained for Kennel Club working trials. He would jump over large obstacles on command and his owner knew about how far and how high he could jump. Jumping was studied because it is a particularly strenuous activity: the forces (and stresses) involved are higher than in trotting or galloping at constant speed.

The basic piece of equipment for the investigation was a force platform. Fig. 3.1 shows in principle how force platforms work. It shows a platform flush with the floor, mounted on springs. When a force acts on the platform (for instance, when someone steps on it) the springs are distorted. The distances x and y change by amounts proportional to the components F_x and F_y of the force. Transducers monitor the distances x and y and give electrical outputs

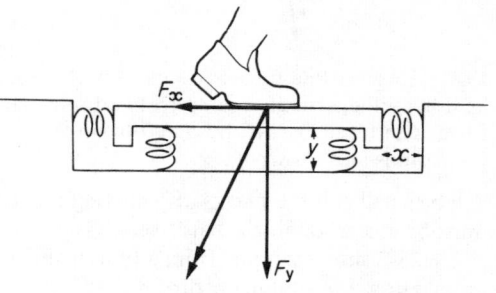

Fig. 3.1 A diagram illustrating the principle of the force platform.

proportional to F_x and F_y, respectively. Additional springs and transducers can be fitted to give the third component of the force, at right angles to the page.

Actual force platforms are not as simple as the one in the diagram, and they are expensive. A major problem in designing them is to make them capable of registering forces which change rapidly. A spring-mounted platform with an animal or other mass on it will tend to vibrate with a particular period which will be called T. It could be damped to prevent the vibrations but even if the amount of damping was ideal it would require a time about equal to T to register a change of force. It could not follow satisfactorily changes of force occurring in times shorter than T. Hence T must be made short. This means that the springs (or their equivalent) must be very stiff, particularly if the combined mass of the platform and animal is large. If the springs are very stiff the changes in x and y

17

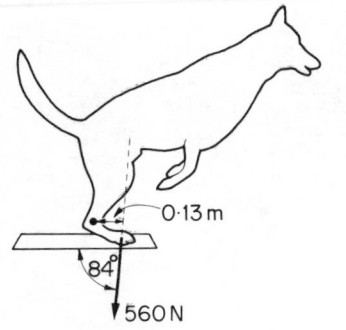

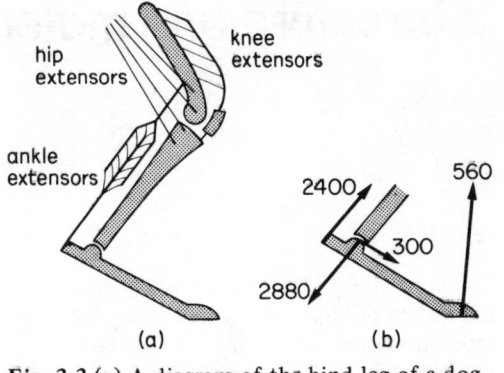

Fig. 3.2 An outline traced from a film of a 36 kg dog taking a running jump from a force platform. The force exerted by one foot on the platform is also shown.

(a)   (b)

Fig. 3.3 (a) A diagram of the hind leg of a dog showing the main groups of muscles involved in jumping. (b) Forces acting on the foot at the instant shown in Fig. 3.2.

will be small and the means of detecting them must be correspondingly sensitive.

Fig. 3.2 shows a frame from a film of the trained dog. He is taking off from the force platform to jump over a row of hurdles. At this particular instant there are two hind paws on the platform and since they are symmetrically placed it seems reasonable to assume that they are exerting equal forces. The record shows that at this instant the total force on the platform was 1120 N at 84° to the horizontal, so it can be assumed that each paw exerted 560 N as shown. The line of action of this force is at a distance of 0·13 m from the ankle joint so the moment about the joint is $0·13 \times 560 = 73$ N m.

This moment must be produced by the extensor muscle of the ankle, which are shown in Fig. 3.3a. They act through Achilles tendon which is 30 mm from the axis of the joint so the force they must exert is $73/0·030 = 2400$ N (0·24 tonne wt.). It can be calculated from the cross-sectional area of the tendon that the stress in it must have been 110 MN m^{-2}. The instant being analysed was the instant when the force was greatest, in a very strong jump, but the stress is remarkably high. Measurements of the tensile strength of tendons have generally given

values of 100 MN m^{-2} or less. The dog did not break its Achilles tendon in this or any other jump, and veterinary surgeons tell me that dogs rarely injure their Achilles tendons. It may be that the force which acts briefly in a jump and does no damage, would break the tendon if it were allowed to act for longer (for instance, in a strength test). Nevertheless, it seems unlikely that the tendon has much strength to spare, over and above what is needed for the jump.

The stress needed in the muscles to exert this force can also be calculated, but it is necessary to take account of the arrangement of fibres within the muscles. The hip extensor muscles are parallel-fibred: that is to say, their muscle fibres run lengthwise along them, as indicated in Fig. 3.3a. The knee and ankle extensors are pennate: their fibres run obliquely across them and attach to the sides of tendons. It can be calculated from the dimensions of the ankle extensors and from the angles of their fibres that to produce the calculated force of 2400 N, a stress of 310 kN m^{-2} must act in the fibres. Similarly it can be calculated that the knee extensors must have exerted stresses up to 190 kN m^{-2} and the hip extensors up to 270 kN m^{-2} in the

18

course of take-off for this jump. Experiments with isolated muscles from other (smaller) animals have shown that vertebrate striated muscle can exert stresses of 200–350 kN m^{-2} when maintaining constant length, and less when shortening (see p. 15), so the stresses in the dog muscles are about what might be expected. When dogs are galloping at top speed their muscles contract faster, and exert smaller forces.

The same data can be used to calculate stresses in leg bones. Consider the tibia at the instant represented by Figs. 3.2 and 3.3b. What force is acting on its distal end? By considering the equilibrium of the foot it can be shown that the tibia must be exerting on the foot a force which has components 2880 N parallel to the tibia and 300 N at right angles to it, as shown in Fig. 3.3b. Equal and opposite forces are exerted by the foot on the tibia. The 2880 N component tends to set up compressive stresses in the tibia. The 300 N component tends to bend it, setting up tensile stresses in its anterior face and compressive stresses in its posterior face. The resulting stresses in any particular cross-section of the bone can be calculated, if the dimensions of the bone are known. The formulae required are standard ones, used by engineers. Calculations for the cross-section half way along the bone gave stresses ranging from a tensile stress of 60 MN m^{-2} at the anterior edge to a compressive stress of 100 MN m^{-2} at the posterior edge. The stress at the anterior edge is a tensile one because the bending effect of the 300 N component predominates over the compressive effect of the 2880 N component. The bending moment due to the 300 N component is proportional to the distance from the distal end of the bone, but the bone tapers in such a way as to keep the stresses about the same all along its length.

Bone can withstand considerably higher stresses. In very careful experiments in which bone was kept moist and stretched quickly, the tensile and compressive strengths were found to be 170 MN m^{-2} and 280 MN m^{-2}, respectively (Chapter 2, Ref. [5]). Does this mean that the dog's tibia is about three times as strong as it needs to be? The question is not easily answered. On the one hand the yielding effect described on p. 13 must make the bone even stronger in bending than simple calculations based on tensile strength suggest. On the other hand greater forces may act when the dog falls or has some other accident, than can be exerted in jumping. Consider what happens when a dog takes off for a jump. At the instant when the greatest forces act, the hip and knee extensor muscles are shortening and must be exerting less than the maximum isometric force. The ankle extensor muscles are stretched and then shorten in take-off (p. 32) and at the instant when they exert most force are at their maximum length and probably exerting the maximum isometric force possible at that length. Now consider what happens when the dog lands after a jump or a fall. The joints in its legs bend. The extensor muscles resist the bending but are forcibly stretched, and the forces they exert may be well above their isometric forces (Fig. 2.9). When frog muscle is stretched rapidly it may develop twice the maximum isometric force. The stresses in the leg bones do of course depend on the forces exerted by the muscles which attach to them. If a dog landed from a fall with its leg muscles relaxed, its legs would simply fold up as they hit the ground and would be unlikely to break. Ribs might get broken, but not leg bones.

3.2 Other organisms and other methods

I have written a lot about the experiments with dogs on force platforms because the force platform is a very useful instrument and because the experiments illustrate rather well some of the things that can be done with it. Force platforms have been used in similar ways in experiments with men, kangaroos, frogs and running birds. The stresses found to occur in the bones and muscles of men and kangaroos in strenuous activities are quite similar to the stresses found in dogs. However it was not possible to demonstrate such high stresses in jumping frogs [5] or running quail [6].

Stresses in bones can also be obtained by experiments with strain gauges. These actually measure strain, not stress, but if Young's modulus is known the stress can be calculated. The simplest type of strain gauge is a piece of very fine wire or metal foil glued to paper or some other insulating material. This is glued in turn to the object being investigated. If the wire is stretched it becomes slightly longer and thinner and its electrical resistance increases. If it is made to shorten its resistance decreases. If it is incorporated in a suitable circuit these changes can be detected. More sensitive strain gauges are made from semiconductor materials. A single strain gauge only registers stretching and compression in a single direction, but groups of gauges running in different directions can be used.

In some recent experiments strain gauges have been glued to bones in living sheep and dogs [9]. This involves a surgical operation, and the bone surface has to be cleaned carefully if the gauges are to adhere firmly. The strain gauges in the animal were connected through long wires to recording equipment, and the animals were free to walk about. Strains up to about 0·03% were recorded from the surfaces of the vertebrae, tibia and calcaneus of sheep, as they walked and trotted. Since Young's modulus for fresh wet bone is about 10–20 GN m^{-2} these correspond to stresses up to about 3–6 MN m^{-2}. This is far less than the stresses calculated from the force platform records of dogs jumping, but higher stresses would presumably have acted in more strenuous activities.

Some muscular activities produce pressure differences, so that pressure measurements can be used to calculate muscle stresses. For instance, most teleost fish feed by sucking food into their mouths. The pressure changes involved are very rapid, which makes them a little awkward to record: the sucking action may last as little as 30 ms. Ordinary manometers cannot respond fast enough to register pressure changes like these but electrical pressure transducers can, provided proper precautions are taken. Any tube leading from the transducer to the animal must be reasonably rigid, and filled with a liquid free from air bubbles. A popular type of pressure transducer contains a diaphragm with strain gauges bonded to it. The diaphragm is distorted by the pressure and the distortions are detected by the strain gauges.

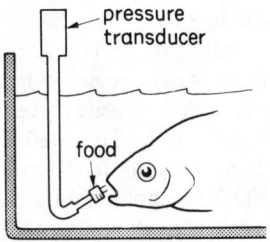

Fig. 3.4 An experiment in which pressures were recorded from the mouths of teleosts as they fed.

Fig. 3.4 shows how a pressure transducer was used to record pressures inside the mouths of teleosts during feeding [1], [2]. A tube from the transducer dips into the aquarium. Over its end is slipped a ring of body wall cut from an earthworm. The fish learns to take this food by putting its mouth round it and sucking. Films show that the movements involved are almost or exactly the same as in normal feeding. Some sucks were stronger than others and some species could apparently suck harder than others. The maximum pressure difference recorded between the mouth and the surrounding water varied between species between about 0·8 and 4 m water (0·08 and 0·4 atm). Measurements of skull and muscle dimensions were made on three species and used to calculate the muscle stresses needed to produce the maximum pressure differences. Stresses of 250, 300 and 370 kN m^{-2} were calculated for one of the principal muscles involved, in the three species. They lie in the range of values one would expect for maximum isometric stresses.

An attempt has been made to calculate the stresses exerted by the flight muscles, when pigeons fly [3]. The power needed for flight at top speed has been estimated as 13 W. The wings beat 5·6 times per second so the work done in each beat is $13/5·6 = 2·3$ J. The main wing muscles shorten by 2 cm (2×10^{-2} m) in each beat, so they must exert a total force of about $2·3/2 \times 10^{-2} = 120$ N. It can be calculated from this and from the dimensions of the muscles that a stress of 90 kN m^{-2} required. This is far less than the likely maximum isometric stress, but it should be about right for maximum power output (p. 15).

If the muscles had too low a mechanical advantage they would contract relatively slowly in flight at a given speed and have to exert relatively large forces. If they had too high a mechanical advantage they would contract faster and exert smaller forces. In the first case top speed would be reached while the stress in the muscles was too large for maximum power output. In the second, it would be reached while the stress was too low. It is only if the mechanical advantage is nicely adjusted that the full capacity of the muscles as a source of power can be used at top speed. If it is so adjusted, as it seems to be, the bird can fly faster than would otherwise be possible.

So far this chapter has been about bone, muscle and tendon. Consider now the resilin which serves as a catapult when fleas jump (Chapter 2, Ref. [3]). Films of fleas jumping show them leaving the ground at 1 m s^{-1}. The masses of these particular fleas were about 0.5 mg (5×10^{-7} kg) so the kinetic energy at take-off was about $\frac{1}{2} \times 5 \times 10^{-7} \times (1)^2 = 2·5 \times 10^{-7}$ J. The fleas have plenty of muscle to provide this amount of energy in a single contraction, but it has to be provided very fast. The fleas' legs are only about 0·4 mm long, so acceleration to 1 m s^{-1} has to be achieved in this short distance. The mean velocity, in acceleration from rest to 1 m s^{-1} is 500 mm s^{-1} so the 0·4 mm is covered in $0·4/500 = 8 \times 10^{-4}$ s (0·8 ms). Producing 2.5×10^{-7} J in this time demands a power output of 3×10^{-4} W. This may not seem much, but it is far more than the muscles in a tiny flea can be expected to produce. This is why a catapult is needed.

A boy using a catapult uses his muscles to stretch the rubber. He stretches it relatively slowly, storing energy which is released very rapidly when he lets the rubber go. He exerts a relatively small power over a relatively long time and the catapult exerts a larger power for a shorter time.

The flea catapult does essentially the same thing. It is stressed relatively slowly by a muscle and then released by a trip mechanism. It recoils rapidly, extending the legs and releasing far more power than the muscle could, though for a very short time. A catapult can only be stretched so far or it will break. To store a given amount of energy and fire a stone of given mass at given velocity, it must contain a certain minimum volume of rubber. To store the $2·5 \times 10^{-7}$ J needed for the fleas' jump, a certain minimum volume of resilin is needed. It can be calculated from the volume of resilin in the flea and from Young's modulus that when the catapult is set, ready for the jump, the stress in the resilin must be about 2 MN m^{-2}. The tensile strength of resilin is about 3 MN m^{-2} so there seems to be an adequate reserve of strength.

The branches of plants have to withstand wind forces, and have to support their own weight. The forces exerted by winds of various speeds on young conifer trees, 8 m high, have been measured in a large wind tunnel [8]. It was found that winds of about 18 ms^{-1} (40 mph) exerted forces about equal to the weight of the tree (excluding roots). Such wind speeds occur in gales, and gusts of wind may be much faster. Trees in exposed situations are liable to suffer wind forces equal to or greater than their weight. The same is presumably also true of single branches.

It is not easy to measure the maximum wind forces which are likely to act in a particular situation, but it is easy enough to weigh a cut branch. It is therefore much easier to calculate stresses due to weight than stresses due to wind. Can we identify branches in which weight stresses are likely to predominate and wind stresses to be unimportant? Obviously we should choose branches in sheltered situations, for instance in woods. We should also choose horizontal branches, for the following reason. It is easier to break a branch, or any other long thin object, by bending it than by pulling or pushing on its ends. Wind forces act horizontally and so can exert large bending moments both on vertical and on horizontal branches. Weight acts vertically, and so will exert larger bending moments on horizontal branches than on more vertical ones of similar dimensions.

A horizontal branch of holly (*Ilex aquifolium*) was cut in the interior of a small wood. It was 4·0 m long and its centre of mass (located by balancing) was 1·8 m from the cut end. Its mass was 3·2 kg so its weight must have exerted a bending moment of $3·2 \times 9·8 \times 1·8 = 56$ N m at the cut end. At this end it was cylindrical and its diameter (without bark) was 32 mm. It can be calculated from standard formulae (see, for instance, [10]) that the weight must have set up a tensile stress of 17 MN m^{-2} at the top of the branch and a compressive stress of 17 MN m^{-2} at the bottom. Similar calculations for other small branches of various species have given similar results.

These stresses are quite small. The tensile strength of wood is about 200 MN m^{-2}, wet or dry.

3.3 Pennate muscles and tubular skeletons

The ankle and knee extensor muscles of the dog are pennate while the hip extensor muscles are parallel-fibred (Fig. 3.3a). Why should this be? What are the special merits of pennate and parallel-fibred muscles?

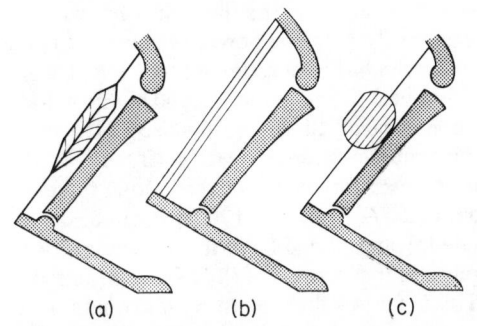

Fig. 3.5 Diagrams of the leg skeleton and an ankle extensor muscle of a real dog (a), and of two imaginary dogs which are discussed in the text.

As a general rule, pennate muscles contain more muscle fibres than parallel-fibred muscles of the same shape, but each fibre is shorter. Consequently the pennate muscle can exert more force than the parallel-fibred one but cannot shorten so much. Fig. 3.5a shows one of the ankle extensor muscles of a dog. It is pennate, with very short muscle fibres. Dogs could have evolved a parallel-fibred muscle which would have had exactly the same effect on the ankle joint. A very short, fat parallel-fibred muscle could have exerted the same force and shortened by the same amount. It would, however, have made an awkward bulge on the leg (Fig. 3.5c). Alternatively a long, slender parallel-fibred muscle might have evolved. It could not have exerted as much force as the pennate muscle and it could have shortened more, but if it had a long enough lever arm it would exert the same moment about the ankle and move it through the same angle. For the lever arm to be long enough, the heel would have to be very long (Fig. 3.5b). There would be an awkward gap between the muscle and the tibia. The pennate muscle (Fig. 3.5a) allows a much neater arrangement than either of the parallel-fibred muscles (b and c). In other positions parallel-fibred muscles may fit more neatly. The parallel-fibred hip extensor muscles could be replaced by pennate muscles with shorter lever arms, were it not for the geometrical difficulty of fitting in a sufficient bulk of muscle close enough to the femur to have the shorter moment arm.

Insects and other arthropods have slender limbs so the lever arms are necessarily short. The limb muscles must exert correspondingly large forces and most of them are, appropriately, pennate.

The major bones of mammal legs are hollow tubes, not solid rods. Does this give an advantage? It probably does, for a tube is a good shape for a structure which has to resist bending moments and yet be as light as possible. Bicycle frames and scaffolding are made tubular rather than solid for this very reason. When a rod is bent, one side is stretched and the other compressed. The material near the middle is stretched or compressed very little and so contributes little to resisting the bending moment. Material from the middle of the rod can be removed, making the rod a tube, without reducing the stiffness or the strength in bending much. Similarly a tube is stiffer and stronger in bending than a solid rod of the same weight and length.

It is convenient to define a quantity k, which is the ratio of the internal diameter of a tube to its external diameter. For a solid rod $k = 0$. A stout tube with a relatively thin wall has a high value of k and if the wall is infinitesimally thin $k = 1$. The tibia of a dog is a tube with $k \simeq 0.5$ and it can be calculated from standard engineering formulae that it contains only about 0.8 times as much bone as a solid rod of the same length which could withstand

the same bending forces. More bone could be saved if the value of k were increased. If it had $k = 0.9$ like the humerus of a swan, it would only require 0.4 times as much bone as the equivalent solid rod.

It is generally advantageous for limbs to be as light as possible, to keep the energy needed for locomotion as low as possible (p. 29). The dog tibia is filled with marrow, which is about half as dense as bone. It can be calculated that it will have minimum weight (including the weight of marrow) for given strength if $k = 0.4$. If k were bigger less bone would be needed but a greater weight of marrow would be needed to fill the cavity and the total weight would be greater. The dog tibia seems to have nearly ideal proportions, for a marrow-filled bone.

The swan humerus is filled with air. Air-filled bones can with advantage have larger values of k than marrow-filled bones, but there is a limit. A tube with too high a value of k does not break but fails by kinking, as a plastic drinking straw kinks when it is bent. Once this point is reached further increase in k increases the weight of material needed for given strength. The ideal value of k depends on the properties of the material. The swan humerus might be in danger of kinking if it were a simple tube, but it has internal struts which must help to prevent kinking.

Arthropods have exoskeletons. The limb skeleton is tubular, with the muscles inside. The muscles may be heavy, but they are needed in the leg anyhow. There is no need to load the leg with marrow or any other inert space-filler and k can be high. It might be possible to make a dog's leg lighter without loss of bending strength, by providing it with an exoskeleton. However, this exoskeleton would be liable to damage by impact [7].

A catapult made of rubber or resilin can only store a limited amount of elastic energy without breaking (p. 21). Conversely, a certain minimum amount of energy is needed to break a catapult, a skeleton or a china cup. A falling cup gains

23

kinetic energy. It must be dropped from a certain minimum height to have enough kinetic energy, when it hits the floor, to break. The energy needed to break a skeleton of a given material is proportional to its bulk. For animals with equal proportions of skeleton in their bodies, the energy the skeleton can absorb in an impact is proportional to the mass m of the body. Let this energy be Km. Suppose an animal of mass m running with velocity u collides with an immovable object. It is brought to rest, losing kinetic energy $\frac{1}{2} mu^2$. If u is large enough this will be greater than Km and something must break, irrespective of the value of m. Large animals are generally faster than small ones and so are in more danger of smashing their skeletons. An unprotected exoskeleton may be fine for an insect, but it would be hazardous for a large mammal. The skin and flesh which cover the bones of mammals absorb some of the energy in impacts, just as a carpet may absorb energy when a cup falls on it and so reduce the danger of breakage.

The stems of grasses and many other plants are tubes, so less material is needed for given strength or stiffness than if they were solid.

References

[1] Alexander, R.McN. (1969), *J. Zool., Lond.,* **159**, 1–15.
[2] Alexander, R.McN. (1970), *J. Zool., Lond.,* **162**, 145–156.
[3] Alexander, R.McN. (1973), In Bolis, L., Schmidt-Nielsen, K. and Maddrell, S.H.P. (eds.) *Comparative Physiology,* 1–21, North-Holland, Amsterdam.
[4] Alexander, R.McN. (1974), *J. Zool., Lond.,* **173**, 549–573.
[5] Calow, L.J. and Alexander, R.McN. (1973), *J. Zool. Lond.,* **171**, 293–321.
[6] Clark, J. and Alexander, R.McN. (1975), *J. Zool. Lond.,* In press.
[7] Currey, J.D. (1967), *J. Morph.,* **123**, 1–16.
[8] Fraser, A.I. (1962), *Rep. Forest Res.,* 178–83.
[9] Lanyon, L.E. (1973), *J. Biomechanics,* **6**, 41–9.
[10] Warnock, F.V. and Benham, P.P. (1965), *Mechanics of Solids and Strength of Materials,* Pitman, London.

4 Animal locomotion

4.1 Power requirements

The mechanisms of most types of animal loco-motion are well understood but there is a great deal of research in progress on the power needed for locomotion. This section of the chapter is about measurements of the power which is used. Subsequent sections describe investigations of the reasons why these amounts of power are required.

The most generally useful method of measuring the power an animal is using is to measure the rate at which it is using oxygen. This is of course only feasible if the animal is releasing the energy it needs by aerobic metabolism and not building up an oxygen debt. If an animal keeps going for a long time at the same speed one can generally be reasonably confident that at least most of the energy is being released by aerobic metabolism, since an oxygen debt cannot be allowed to build up indefinitely. Metabolism involving 1 cm^3 oxygen releases about 20 J chemical energy, whatever food is being metabolized.

It is obviously easiest to measure an animal's oxygen consumption if it keeps still, but locomotion involves movement. Many methods of measuring oxygen consumption in locomotion involve moving the environment so that the animal remains stationary.

Fig. 4.1a shows a terrestrial animal running on a conveyor belt. It is trained to run so as to remain stationary while the belt moves. If it stops running it is carried back against a barrier or falls off the end of the belt. This system has

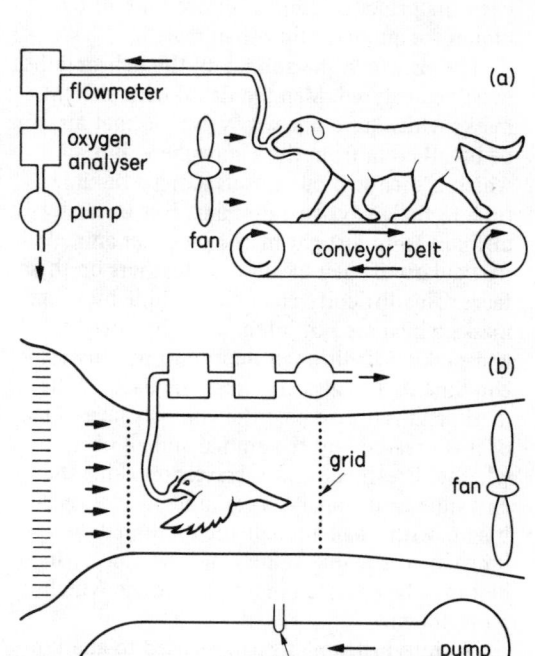

Fig. 4.1 Three experiments in which oxygen consumption during locomotion is measured. In each case the animal remains stationary while the environment moves.

been used a great deal for experiments with man, but a remarkable range of animals have been trained to run on conveyor belts of suitable size and speed. They include lizards, rheas, mice, cheetahs, kangaroos and many others [5], [14], [15], [16]. A kangaroo hopping on a conveyor belt is a particularly intriguing sight, bouncing vertically up and down while the belt moves under it. Knut Schmidt-Nielsen and Richard Taylor and their colleagues at Duke and Harvard Universities, respectively, have been the principal users of conveyor belt techniques for animals other than man.

The air which the animal on the belt breathes must be analysed. Men are fitted with airtight masks which have one-way valves so that air can be breathed in from the atmosphere, but air which is breathed out travels along a flexible tube to the analysis equipment. It is generally difficult to fit airtight masks on other animals, particularly if they have fur or feathers on their faces. The difficulty can be overcome by using masks which are not intended to be airtight, and sucking air through them into the analysing equipment. The air leaks in at the back of the mask, and is drawn past the animal's face. Some of it is breathed by the animal and all of it, whether breathed or not, travels on along the exit tube to the analysis equipment. In experiments with small animals the mask may be omitted: the whole animal and the conveyor belt may be enclosed in a box through which air is drawn.

Various instruments can be used to analyse the air. One of the most convenient is the paramagnetic oxygen analyser, which determines the partial pressure of oxygen in air passing through it. If the rate of flow of air is also measured the rate at which oxygen is being used can be calculated.

The fan shown in Fig. 4.1a blows air past the animal at the speed of the conveyor belt. This makes conditions as nearly as possible the same as if the animal were running in still air on stationary ground. This refinement of the technique is often unnecessary because power expended in overcoming air resistance is generally only a small fraction of the total power, at least at the relatively low speeds at which no oxygen debt builds up (see [13]).

Fig. 4.1b shows how Vance Tucker of Duke University performs the equivalent experiment with birds [19], [20]. The bird is flying against the air flow in a wind tunnel, keeping stationary relative to the ground. The tunnel is big enough for a man to get in beside the bird, while it is being trained. If it lands on the floor he picks it up and tosses it into the air again. Budgerigars and gulls have been trained to fly in the tunnel and their oxygen consumption has been measured by the leaky mask technique.

If relative movement between air and bird is to be the same as if the bird were flying in still air, all the air must approach the bird at the same velocity. This is not easy to achieve. Air tends to flow faster at the centre of a tunnel than near its walls. Air flowing fast in a large channel quickly becomes turbulent: that is to say, it becomes a mass of swirling eddies. The tunnel is designed to overcome these problems as well as possible.

Fig. 4.1c shows equipment which has been used by J.R. Brett and by others to perform essentially the same experiment with fish [2]. The fish swims against the current in a water tunnel which is designed on the same principles as wind tunnels. If it stops swimming it is carried back against a grid which can be mildly electrified during training to encourage it to keep swimming. The tunnel is a complete circuit and the water travels round it repeatedly. Its oxygen content falls progressively as the fish uses the oxygen up. It must not be allowed to fall too far or the fish's performance will be affected. The declining oxygen concentration is measured by means of the oxygen electrode. If the total volume of water in the system is known the rate at which oxygen is being used can be calculated.

The water tunnel must give the fish plenty of room to swim. A really large water tunnel would

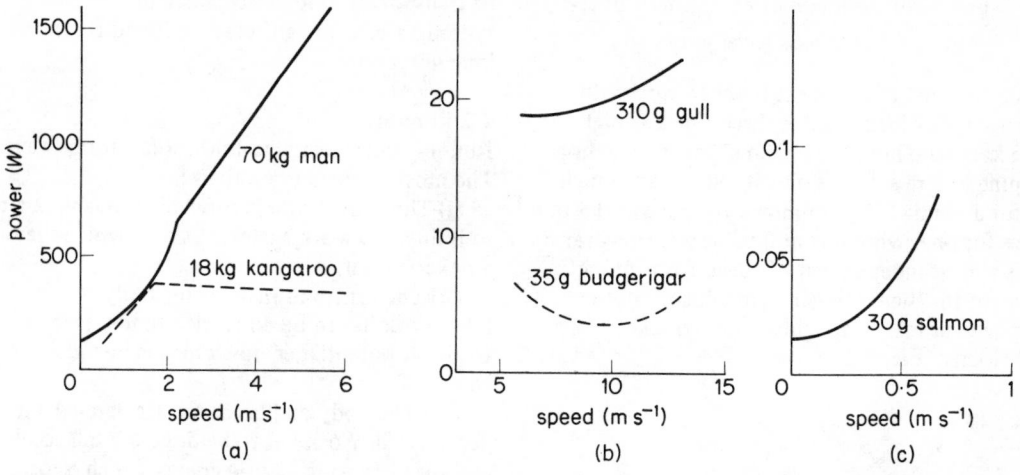

Fig. 4.2 Graphs of power requirement (calculated from oxygen consumption) against speed for (a) terrestrial locomotion (b) flight and (c) swimming. Data from References [2], [3], [5], [19], [20].

be a major piece of engineering, but tunnels already in use will accommodate fish up to about 300 g. Pacific salmon, trout, haddock and a few other species have been used.

Fig. 4.2 shows some results from the experiments illustrated in Fig. 4.1. Each graph shows power plotted against speed. The graphs for terrestrial locomotion have breaks in them where the gait changes. Men run at high speeds and walk at low speeds. Kangaroos hop at high speeds, and at low speeds use the pentapedal gait which involves all four limbs and the tail. Both men and kangaroos automatically choose the gait which requires least power at the speed in question. Extrapolation of the graphs indicates that walking would require more power than running at speeds above about 2·3 m s^{-1}, and running would require more power than walking at lower speeds. Notice that once the kangaroo is hopping, further increase in speed requires no extra power. Indeed slightly less power seems to be used at 6 m s^{-1} than at 2 m s^{-1}.

Fig. 4.2b shows that a budgerigar needs less power to fly at about 10 m s^{-1} than to fly either faster or slower. However, the power used by the gull increases with speed, over the whole range of speeds. The gull used more power than the budgerigar because it was larger, but though it was nine times the weight of the budgerigar it only used about five times as much power.

Fig. 4.2c shows that the power used by a fish increases rapidly as speed increases.

Subsequent sections of this chapter show how far the data of Fig. 4.2 can be explained. The rest of this section makes comparisons between running, flight and swimming, and between animals of different size. It will be convenient to use a quantity called the net cost of transport, which is the energy needed to move a unit mass of animal a unit distance. An animal of course uses oxygen when resting and presumably has to go on using oxygen at the same rate, for purposes other than locomotion, when it is moving. The net cost of transport is therefore calculated as:-

$$\frac{\text{power during locomotion} - \text{power at rest}}{\text{body mass} \times \text{velocity}}$$

Net cost of transport is apt to vary with speed. For instance it is three times as high for a kangaroo hopping at 2 m s^{-1} as for one hopping at 6 m s^{-1}. It is only about half as much for a young (30 g) salmon swimming at 0.5 m s^{-1} as for one swimming at 0.65 m s^{-1}. However it is almost independent of speed for running men: for them (power during locomotion − power at rest) is roughly proportional to velocity (Fig. 4.2a).

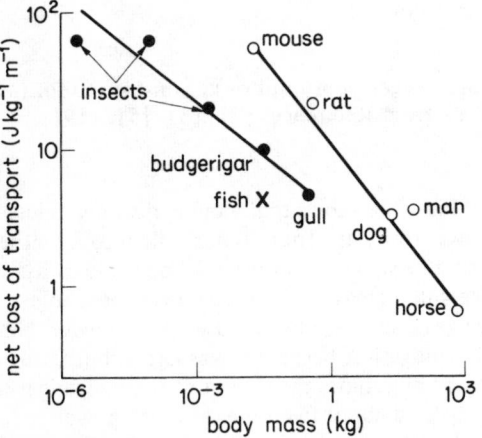

Fig. 4.3 Net cost of transport plotted against body weight for running (○), flight (●) and swimming (×). Note that logarithmic co-ordinates have been used. Data from Refs. [2], [16], [21].

Fig. 4.3 shows relationships between net cost of transport and body weight for running, flying and swimming animals. The cost decreases in a regular way as body weight increases. It is higher for terrestrial locomotion than for either flight or swimming [21]. Flying birds use far more power than running mammals of similar size, but they travel so much faster that the cost of transport is lower. It may be because the cost

of transport is so high that small mammals do not migrate, although some birds and fish make long migrations.

4.2 Running

Running uses mechanical power in various ways. The most important are these:-

(i) The animal moves forward through the air so it must do work against the drag which the air exerts on it.

(ii) The centre of mass of the body rises and falls. Work has to be done each time it is raised to give it potential energy which is lost again when it falls.

(iii) The body and limbs are accelerated and decelerated. Work has to be done in each acceleration to provide kinetic energy which is lost in the subsequent deceleration.

Small amounts of mechanical power are also used to overcome friction in the joints, to pump blood faster than would be necessary while resting, etc.

Let us see how a man uses mechanical energy when he runs. The power needed to overcome drag ((i) above) can be calculated roughly by simple aerodynamics and shown to be quite small, as is also indicated by measurements of oxygen consumption [13]. The fluctuations of potential and kinetic energy ((ii) and (iii)) could in principle be obtained by careful analysis of films. The centre of mass shifts within the body as the limbs move but its height could be calculated for each position adopted in running, so the potential energy changes could be obtained. The fluctuations of velocity of each part of the body could be obtained from the films and used to calculate the kinetic energy changes. In practice it is generally very difficult to make accurate enough measurements on films to determine the potential and kinetic energy changes with even reasonable precision. A much better approach involves the use of a force platform [3], [4].

This approach makes a distinction between

28

two components of kinetic energy. Consider an animal moving in space which we will suppose for the moment to be one-dimensional. The animal consists of a part of mass m_1 which has velocity u_1, another of mass m_2 which has velocity u_2, and so on. The total mass is Σm and the velocity of the centre of mass is $\bar{u}$. The total kinetic energy of the animal is plainly

$$\tfrac{1}{2}(m_1 u_1^2 + m_2 u_2^2 + m_3 u_3^2 + \ldots)$$

It can be shown that this is equal to

$$\tfrac{1}{2}(\Sigma m)\bar{u}^2 +$$
$$+ \tfrac{1}{2}(m_1(\bar{u} - u_1)^2 + m_2(\bar{u} - u_2)^2 + \ldots)$$

The first term, $\tfrac{1}{2}(\Sigma m)\bar{u}^2$, is known as the external kinetic energy. It is the kinetic energy due to movement of the animal as a whole. $\tfrac{1}{2}(m_1(\bar{u} - u_1)^2 + m_2(\bar{u} - u_2)^2 + \ldots)$ is known as the internal kinetic energy. It is the kinetic energy due to movement of the parts of the animal relative to each other. The formulae have been written for movement in one dimension but can easily be extended to cover movement in three.

Fig. 4.4 shows a force platform record of a man running, and how it can be used to determine the changes of potential and external kinetic energy. Each foot exerts a forward force on the ground followed by a backward one. The ground exerts equal, opposite forces on the foot: first a backward force which decelerates the body and then a forward one which accelerates it. One form of Newton's second Law of Motion is that the force on a body is equal to the rate at which the momentum of the body changes. Hence impulse (which is the time integral of force) equals total change of momentum. The area C (Fig. 4.4) represents an impulse of 11 Ns which must reduce the forward momentum of the body by 11 kg m s^{-1}. Since the mass is 68 kg the velocity of his centre of mass must have been reduced by $11/68 = 0.16$ m s^{-1}. His mean velocity is known from a film so the

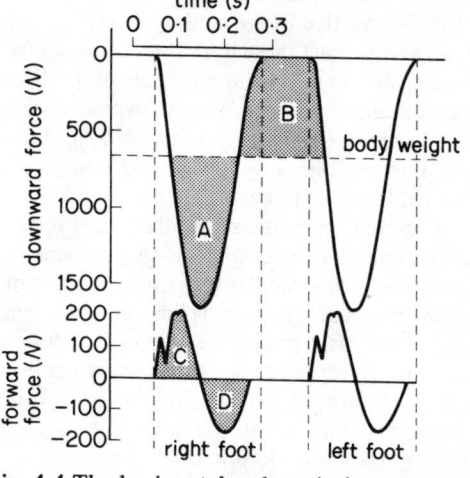

Fig. 4.4 The horizontal and vertical components of the force exerted on the ground by the feet of a 68 kg man running at 3.5 m s^{-1}. The forces exerted by the right foot were actually recorded. The same forces have been appropriately displaced in time to represent the forces which the left foot (which did not land on the platform) presumably exerted.

reduction in external kinetic energy can be calculated. D has the same area as C and represents an equal but opposite change of momentum. The net upward force on the body is the force registered by the platform minus the force exerted on the body by gravity. Hence area A represents an upward impulse and area B (which is equal to it) a downward one. These areas can be used to calculate the fluctuations of the vertical component of the velocity and of the corresponding component of the kinetic energy. Further calculation shows how much these impulses make the centre of mass rise and fall, so the changes of potential energy can be obtained.

The fluctuations of internal kinetic energy cannot be obtained from the force platform record. They have to be obtained by analysis of film. This can be done reasonably accurately because most of the internal kinetic energy is

29

associated with parts of the body which move quite fast relative to the centre of mass.

Analysis along these lines shows that a man running at 6 m s^{-1} (13 m.p.h.) uses about 8% of his mechanical power output to overcome drag, 20% to provide potential energy, 40% to provide vide external kinetic energy and 32% to provide internal kinetic energy [4], [13].

A swinging pendulum repeatedly converts kinetic energy to potential energy and back again. Something similar happens when a man walks. The centre of mass is highest and moving most slowly as it passes over the supporting foot. It is lowest and moving fastest when both feet are on the ground and the weight is being transferred from one to the other. Interconversion of kinetic and potential energy is not possible in running because kinetic and potential energy rise and fall together. The centre of mass is *lowest* and travelling most slowly as it passes over the supporting foot.

This analysis may leave the impression that running wastes a lot of energy. Energy must plainly be used to do work against drag.(unless the man runs with the wind behind him). It must also be used to provide internal kinetic energy for a man cannot run without moving his legs relative to his centre of mass. At first sight it seems wasteful to accelerate and decelerate the body as a whole, causing fluctuations of external kinetic energy. It also seems wasteful to make the centre of mass rise and fall so that work has to be done providing potential energy.

The fluctuations of external kinetic energy are caused by the horizontal components of the force on the ground. Suppose a man ran so that the force on the ground was always vertical. The force on his foot would exert moments about the centre of mass (Fig. 4.5a). The moment would act anticlockwise (as seen in Fig. 4.5) while the foot was in front of the centre of mass and clockwise while the foot was behind it. The man's trunk would pitch (rock forward and back) as he ran. This can be avoided if the

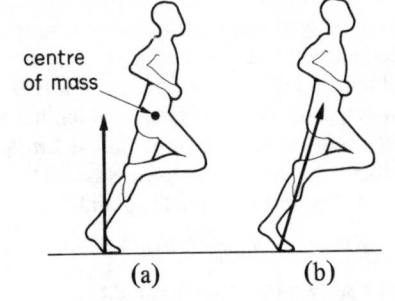

Fig. 4.5 Two possible techniques of running, which are explained in the text.

force on the foot is given a horizontal component, so that it is always more or less in line with the centre of mass (Fig. 4.5b). This is what happens.

Fluctuations of potential energy occur if the vertical component of the force on the ground fluctuates. They must occur in running because there are floating phases when neither foot is on the ground.

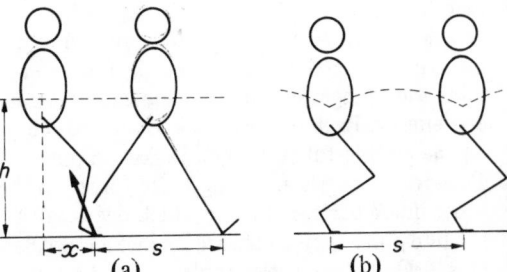

Fig. 4.6 These diagrams are explained in the text.

Fig. 4.6 shows two possible modes of locomotion. A man would look odd using either of them, but they will serve to illustrate how a gait which involves fluctuations of potential energy (such as human running) may require less power than one which does not. Each diagram shows a biped of mass m taking steps of length s and travelling with velocity u. (a) is more like a racing walk than a normal walk. Each foot is placed on the ground as the other is lifted. The vertical component of the force on the ground is kept constant and equal to mg so that the centre of mass remains at a constant height h. The horizontal component is varied so as to keep the resultant in line with the centre of mass, as in Fig. 4.5b. When the centre of mass is at a distance x from a vertical line through the foot the horizontal component must be mgx/h. As the centre of mass moves forward through a small distance δx the horizontal component must change the external kinetic energy by $mgx.\delta x/h$. During a step x changes from $-s/2$ to $+s/2$, and it can be shown by integration that the external kinetic energy must diminish and then increase by

$$mgs^2/8h$$

In Fig. 4.6b contact with the ground is very brief and occurs as the centre of mass passes over the foot. There is no need for any horizontal component in the force on the ground, or any fluctuations of external kinetic energy. However each step is a little jump in which the centre of mass rises and falls through a height Δh. Each jump lasts for a time s/u so the centre of mass rises for $s/2u$ and falls for $s/2u$. An object falling under the influence of gravity falls a distance $gs^2/8u^2$ in this time. The potential energy increases and then diminishes in each jump by $mg.\Delta h$ or

$$mg^2s^2/8u^2$$

Thus the man in Fig. 4.6b uses less energy in each step than the one shown in Fig. 4.6a if

$$mg^2s^2/8u^2 \; < \; mgs^2/8h$$

$$u > \sqrt{gh}$$

For a man, $h \doteq 1$ m and since $g = 10$ m s^{-2} the jumping style of locomotion is the more economical one at speeds above $\sqrt{10} \doteq 3$ m s^{-1} (6 m.p.h.).

Since the styles of locomotion shown in Fig. 4.6 are so unrealistic this result should not be taken seriously. It does, however, make the point that gaits involving periods whn there are no feet on the ground are likely to be advantageous at high speeds. Such gaits include human running, kangaroo hopping and galloping.

Kangaroo hopping is of course rather like the style of locomotion shown in Fig. 4.6b. The length of each hop is roughly proportional to the speed [5] : $s \propto u$ so the duration of each jump (s/u) is constant and the potential energy change in each jump $(mg^2s^2/8u^2)$ is also constant. This helps to explain why the power used does not increase with speed (Fig. 4.2a). However, it is by no means the whole story since the external and internal kinetic energy both fluctuate in the course of a hop. External kinetic energy changes are large and increase with speed, but more energy is saved by elasticity at high speeds.

Most terrestrial animals run on four legs or (in the case of arthropods) more. This section has been about bipedal animals because extra legs make the theory of running harder to handle. Theories of bipedal locomotion are being developed [1a], and we can hope to see them extended to deal with quadrupedal locomotion in the near future.

4.3 Energy saved by elasticity
A child on a pogo-stick bounces along in the manner shown in Fig. 4.6b. Each time he lands energy is stored in a spring, to be released in the elastic recoil as he takes off again. This suggests that energy might be saved by springs built into the body. If there are no springs the kinetic and potential energy lost in each step must be absorbed by muscles which resist the movement of the body. The energy must be dissipated as heat, and replaced by energy provided afresh by active contraction of muscles. If there are springs they can store the energy and release it

31

again when required. The animal can bounce along, using less power. There is evidence that a good deal of energy is in fact saved in mammal locomotion by elastic structures.

Consider human running. The methods described in the previous section show that when a man runs at 6 m s^{-1}, potential and external and internal kinetic energy totalling 200 J have to be provided at each step. Since 3 steps are taken per second the mechanical power required is 600 W. Measurements of oxygen consumption show that the chemical power consumed is • 1500 W. Hence the efficiency of the muscles seems to be 40%. The efficiency of human muscle has been investigated in other experiments. The oxygen consumption of men running uphill on conveyor belts has been compared with level running, and the oxygen consumption of men pedalling stationary bicycles against calibrated brakes has been measured. Both types of experiments show efficiencies of 25% or less. This makes one suspect that the mechanical work required from the muscles in a running step is actually less than 200 J: some of the potential and kinetic energy lost in each step may be stored in elastic structures for reuse.

There is more direct evidence that energy can be stored by elastic structures in human legs [17]. The oxygen consumption of men was measured while they performed a simple exercise, bending their knees and standing straight again twenty times per minute. The knee extensor muscles are tense while the knees bend, controlling the movement. They are stretched as the knees bend and contract to straighten them. In some of the experiments the subjects bent their knees and quickly straightened them again, taking advantage of any elastic recoil of the knee extensor muscles. In others they squatted for a second with knees bent to allow the tension in these muscles to fall and avoid any rebound effect. On average the subjects used $1 \cdot 5$ l oxygen min^{-1} in the rebound experiments

and $1 \cdot 9$ l oxygen min^{-1} when the rebound was avoided. About $0 \cdot 4$ l min^{-1} was used when resting in either a standing or a squatting position. The rebound apparently saved a good deal of energy.

Elasticity probably plays an important energy-saving role in locomotion of many other mammals as well as man. There is evidence that it is important in dogs and kangaroos and this evidence seems to show that the elastic material is not the muscle itself, but collagen.

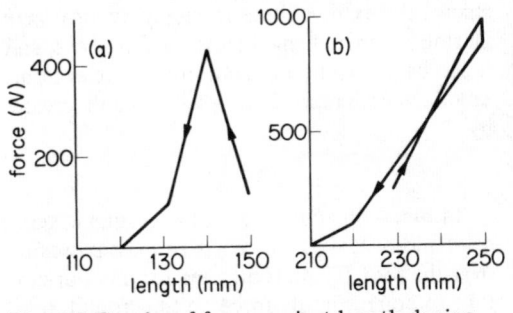

Fig. 4.7 Graphs of force against length during take-off for a jump for (a) the biceps femoris and (b) the gastrocnemius of a 26 kg dog. The biceps is an extensor of the hip and the gastrocnemius an extensor of the ankle. The force platform records do not show how much of the moment about a joint is contributed by each muscle so it has been assumed that equal stresses act in all the hip extensors, and that equal stresses act in both ankle extensors. Based on data in Chapter 3, Ref. [4].

We have seen how the forces exerted by the major leg muscles were calculated from force-platform records of dogs jumping together with film taken simultaneously (p. 18). The lengths of the muscles can also be calculated for each leg position shown in the film. Hence graphs of force against length can be obtained (Fig. 4.7). The biceps femoris shortens throughout the period of contact of the feet with the ground: it does work, accelerating the dog for take-off. The area under the graph represents the amount of work done. The gastrocnemius behaves quite differently. It is forcibly stretched as the foot hits the platform and shortens again as the dog takes off. The force in it increases as it is stretched and diminishes again as it shortens. Work is done on it and then it does about the same amount of work. The graph is essentially the same as is obtained when a spring is stretched and released.

The gastrocnemius is apparently stretched about 3 cm, and then recoils. However, it is a pennate muscle with fibres which, in the resting state, are only 2·5 cm long. A 2·5 cm muscle fibre cannot be stretched by 3 cm and still exert large forces (p. 14). Perhaps the muscle fibres remain more or less constant in length while the long tendons stretch. A calculation using Young's modulus for collagen indicates that the forces which act in jumping are enough to stretch the tendons by about 2·5 cm, so it seems likely that at least most of the stretching occurs in the tendons. The elasticity of the tendons seems to be used in galloping in the same way as in jumping.

It is shown in the next section that elastic structures save energy in insect flight in essentially the same way as in mammalian running.

4.4 Flight

The power needed to propel an aircraft (or a flying animal) has two components [12]. One is the zero-lift power which is used to overcome the viscosity of the air near the surface of the craft, and also provides the energy which is lost as kinetic energy to the air left swirling in the wake. As one would expect it increases as u increases: it is proportional to some power of u between $u^{2 \cdot 5}$ and u^3, and we will take it as being approximately equal to ku^3 where k is a constant for the aircraft in question. The other component of the power is the induced power, which is used accelerating air downwards to provide the upward lift needed to keep the craft airborne. The faster the aircraft goes the more

air it passes and the more air it can accelerate downwards in unit time. The mass of air which can be accelerated in unit time is proportional to u and will be called $k'u$. If it is given a downward velocity w the rate at which the air is given downward momentum is $k'uw$. By Newton's second Law of Motion the force required is $k'uw$. If the aircraft exerts this downward force on the air the air must exert an equal upward force on the craft. To support the craft, this must equal its weight W. Hence

$$W = k'uw$$

$$w = W/k'u$$

The rate at which kinetic energy is given to the air as it is accelerated downwards is $\frac{1}{2}k'uw^2 = W^2/2k'u$. The induced power must equal this. Hence

$$\text{Total power} = \text{zero-lift power} + \text{induced power}$$

$$= ku^3 + (W^2/2k'u)$$

The zero-lift power increases as u increases but the induced power decreases. There is therefore a particular velocity at which the total power is least. For this reason any aeroplane has a particular speed at which its fuel consumption (fuel used per unit time) is a minimum. This explains why the oxygen consumption of a flying budgerigar is less at 10 m s^{-1} than at higher or lower speeds (Fig. 4.2b). It is not clear why there is no minimum in the graph for the gull.

Most of the rest of this section is about gliding and hovering, because recent investigations have thrown a lot of light on the mechanics of these types of flight.

Glider pilots adjust the angle of their craft, so as to glide more or less steeply, by adjusting the elevator flaps. Birds do the same by swinging their wings forward and back. At each angle a range of speeds is possible. Pilots use their air-brakes and birds use their feet to reduce the speed below the maximum for the angle in question.

Consider a glider travelling with a forward speed u and a sinking speed v (the sinking speed is the rate of loss of height). It is losing potential energy at a rate Wv, so power Wv is available to propel it. The power needed (if it is not using its airbrakes) is $ku^3 + (W^2/2k'u)$ (see above). Hence

$$Wv = ku^3 + (W^2/2k'u)$$

$$v = (ku^3/W) + (W/2k'u)$$

The first term on the right hand side of this equation increases as u increases and the second decreases. There is a particular forward speed u at which the sinking speed v can be kept to a minimum. This is the speed at which minimum power is needed for level powered flight. Graphs of sinking speed (without airbrakes) against forward speed for a real glider and a model glider are shown in Fig. 4.8. The former shows the predicted minimum.

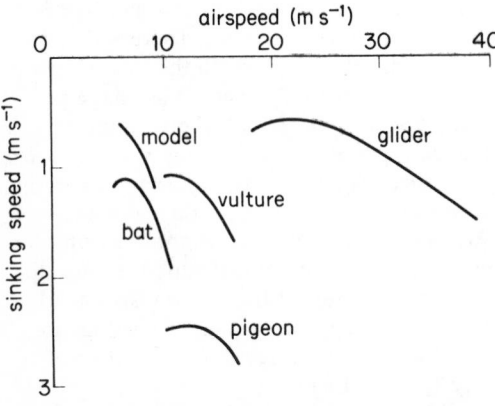

Fig. 4.8 The gliding performance of various birds, a fruit bat, a model glider and a full-sized glider. The minimum sinking speed possible at each forward speed is plotted against the forward speed. All speeds are relative to the air. Data from Refs. [9], [10], [11].

The gliding performance of birds has been observed from the ground and by glider pilots, but the most reliable observations have been made on captive birds gliding in wind tunnels. The data on birds and bats shown in Fig. 4.8 was obtained in this way. The animals were trained to glide so as to remain stationary in a sloping wind tunnel. The wind was kept at a constant speed u while the tunnel was tilted up and down, and the minimum angle θ was found at which the animal could just glide without flapping. The corresponding sinking speed is $u \sin \theta$. The graphs for the animals have the same shape as the graphs for the model and the glider. The vulture, and bat glide almost as well as the model but the pigeon loses height much faster. The glider travels at higher forward speeds than the model or the animals. Because it is bigger, the ratio of its weight to its wing area is greater so it has to travel faster to get enough lift to stay airborne.

Now consider hovering, which is practised by a wide variety of insects and by some small birds and bats. When hummingbirds and insects hover they keep their wings fully extended all the time. This is called normal hovering. Other small birds, and bats, flex their wings at one stage in the stroke. We will consider only normal hovering. The wings beat through a wide angle in a more or less horizontal plane (Fig. 4.9). They work like helicopter blades but move backwards and forwards, not round and round. If a helicopter rotor were reversed it would produce a downward force instead of an upward one but this does not happen in the animals because they change the angle of their wings each time they reverse them. Indeed, they turn their wings upside down.

Torkel Weis-Fogh has shown how the power needed for normal hovering can be calculated [23], [24]. Most animals which hover in this way beat each wing through about 120°. It will be convenient and reasonably realistic to regard the wing movements as simple harmonic motion.

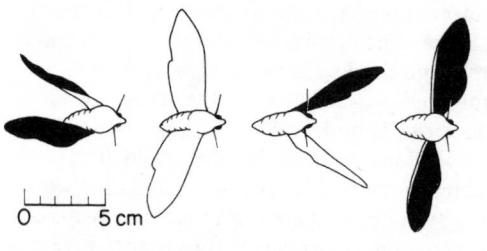

Fig. 4.9 Tracings from film of a moth (*Manduca sexta*) hovering. It is seen vertically from above, and the undersides of the wings are shown black. Successive tracings are separated by intervals of about 12 ms. From Ref. [24].

Consider the moments which must be exerted at the shoulder joint at each stage in the wing beat. Moments are needed to overcome the inertia of the wing, accelerating it at the beginning of each stroke and decelerating it at the end. If the motion is simple harmonic this moment will be proportional to the angle of the wing, taking the mean angle as zero (Fig. 4.10a). The magnitude of the moment can be calculated from the dimensions of the wing and the frequency of the wing beat. Moments are also needed to overcome the aerodynamic drag on the wing which is greatest when the wing is moving fastest in the middle of the stroke (Fig. 4.10b). The magnitude of these moments can be calculated approximately from the dimensions of the wing, the frequency of the beat and the weight of the insect which has to be supported. Measurements of forces acting on insect wings in wind tunnels were useful at this stage in the calculation. Fig. 4.10b is a loop because drag acts backwards in the forward stroke and forward in the backward stroke. Adding the inertial and aerodynamic moments together produces the graph shown in Fig. 4.10c. In the early and middle parts of the forward stroke wing muscles do work against the inertial and aerodynamic moments. They accelerate the wing and overcome drag on the wing, doing the amount of work indicated by the hatched area

in the graph. At the end of the stroke other muscles have to decelerate the wings. They absorb energy (do negative work), and the amount of energy involved is indicated by the stippled area in Fig. 4.10c. Similar amounts of positive and negative work must be done in the backward stroke.

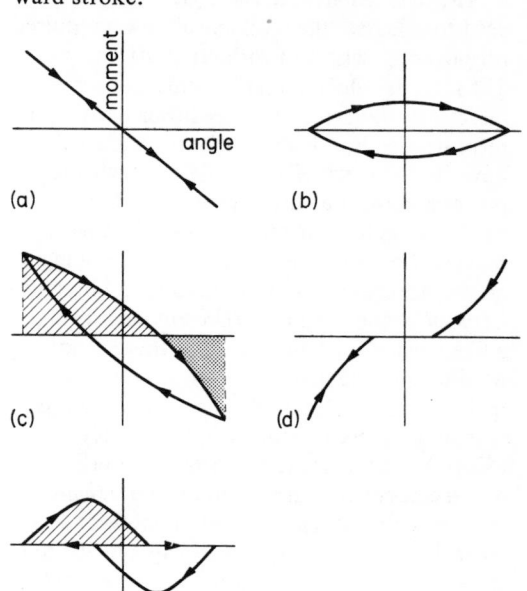

Fig. 4.10 Schematic graphs of moments at the shoulder joint against wing angle in normal hovering. (a) Shows the inertial moment, (b) the aerodynamic moment, (c) the sum of these two, (d) the ideal elastic moment and (e) the sum of (c) and (d). Positive and negative moments act in opposite directions. Further explanation is given in the text.

The amount of work the muscles have to do could be reduced if an elastic structure were incorporated in the shoulder joint, which absorbed energy at the end of each stroke and released it again to help start the next stroke. Ideally it should exert moments as shown in Fig. 4.10d. If these moments acted as well as the inertial

35

and aerodynamic ones, the total moment at each stage in the wing beat would be as shown in Fig. 4.10e. The hatched area in this graph represents the amount of work the muscles would have to do in each forward stroke. It is smaller than the hatched area in Fig. 4.10c.

The hatched areas in these graphs can be used to calculate the mechanical power required for hovering, with and without elastic storage. The rates at which hummingbirds and some insects use oxygen while they hover have been measured, so the chemical power consumed in hovering is known. The calculated mechanical power and the measured chemical power can be used to calculate the efficiency of the flight mucles. The calculated efficiency (assuming no elastic storage) is about 0·2 for a hummingbird (*Amazilia*) and a fruit fly (*Drosophila*). This is a likely value for efficiency, and there is no need to suppose that elastic storage plays a major role in hovering in either case. However, if there is no elastic storage in the hornet (*Vespa*) and the drone fly (*Eristalis*) the efficiencies of their muscles must be 0·6 and 0·5, respectively. This is most unlikely. It seems clear that elastic storage must play a major role in these and many other insects. Insects have structures made of resilin in their thoraxes which could store elastic energy in the required manner, and the cuticle of the thorax could also play a part. The properties of resilin were described in Chapter 2 (p. 9).

There are other exciting investigations of animal flight in progress. One concerns the way in which butterflies and some other insects clap their wings together over their backs as they fly. It is thought that this improves the performance of the wings in flapping flight, by a mechanism unknown in conventional aerodynamics [24].

4.5 Swimming

A flying bird does not merely have to propel itself through the air. It also has to support its own weight, and so it has to expend induced power as well as zero-lift power. A fish which has the same density as the water it is swimming in has no need to support itself. Its power requirement is equivalent to the zero-lift power alone of the bird, and can be expected to be proportional to the cube of the velocity or thereabouts (p. 33). The power used for swimming can be calculated from measurements of oxygen consumption (see for instance Fig. 4.2c). It is not quite proportional to the cube of the velocity and there is another, more puzzling, difficulty. The power requirement is far higher, at any given velocity, than one might expect.

What might one expect? A typical fish is a streamlined, flexible body which propels itself by its own undulations. It is quite different from the rigid bodies studied by engineers but it does not seem too unreasonable to suppose that the power needed to propel it through the water will be about the same as the power needed to propel a rigid body of the same shape at the same velocity.

The rigid body might be driven by a propeller, and the power required would depend on the properties of the propeller as well as of the body. Propellers produce thrust by driving water backwards, so the water left behind in the wake has kinetic energy which can be thought of as energy wasted. Similarly the movements which propel fish give kinetic energy to the water. A mathematical analysis by Sir James Lighthill shows that about 20% of the mechanical work done by the muscles is probably used in this way [8]. This can be expressed by saying that the Froude efficiency is probably about 0·8.

The mechanical power needed to propel a rigid body of the size and shape of a particular fish, using a propeller of this efficiency, can be calculated. The metabolic power which is used is known in some cases from measurements of the oxygen consumption of swimming fish. The ratio of these powers should, one might think, give the efficiency of the swimming muscles.

However it proves to be very low; it seems always to be well under 0·1 [1, 22]. Either the muscles are remarkably inefficient or the mechanical power which is needed is much more than would be needed to propel a rigid body.

The latter seems likely. It may well be quite inappropriate to compare a fish to a rigid body. Most of the power needed to propel a streamlined body serves to overcome the viscosity of the fluid in the boundary layer. This is the thin layer of fluid close to the body which moves with the body. The innermost part of the boundary layer has the same velocity as the body while the fluid outside the boundary layer is stationary, if the body is moving through stationary fluid. Hence there is a steep gradient of velocity in the layer. The thinner the layer the steeper the gradient and the more power is needed to overcome viscosity in the layer. The movements of swimming may make the boundary layer thinner than it would otherwise be, and so increase the power needed for swimming.

This possibility has been investigated [22]. To the backs of fish were attached small plates, standing up at right angles to the long axis of the body. The mechanical power needed to propel these plates at any particular velocity could be calculated, and was checked by direct measurement. The oxygen consumption of fish with plates was compared to the oxygen consumption of similar fish without plates, swimming at the same speed in a water tunnel. Hence the extra power used by the fish to propel the plate could be calculated. The results indicate that if the Froude efficiency is 0·8 the efficiency of the muscles must be 0·1 to 0·2, depending on the speed of swimming. This range of efficiencies is generally accepted as normal for muscle. It seems very likely that the high power consumption of fish swimming is not due to inefficient muscles, but to th·nning of the boundary layer.

Animals use many other mechanisms of locomotion, as well as the ones which have been considered in this chapter. Snails crawl [7], worms and many molluscs burrow [8], squids travel by jet propulsion [6], and so on. The mechanics of these and other forms of locomotion have been and are being studied.

References
[1] Alexander, R.McN. (1974), *Functional Design in Fishes*, ed. 3, Hutchinson, London.
[1a] Alexander, R.McN. (1975), In P.S. Davies (ed.) *Experimental Animal Biology*. In press. Pergamon, Oxford.
[2] Brett, J.R. (1965), *J. Fish. Res. Bd Can.*, **22**, 1491–601.
[3] Cavagna, G.A. (1969), *J. Physiol.*, Paris, **61**, suppl. 1, 3–42.
[4] Cavagna, G.A. Saibene, F.P. and Margaria, R. (1964), *J. appl. Physiol.*, **19**, 249–56.
[5] Dawson, W.R. and Taylor, C.R. (1973), *Nature,* Lond., **246**, 313–4.
[6] Johnson, W., Soden, P.D. and Trueman, E.R. (1972), *J. exp. Biol.*, **56**, 155–65.
[7] Jones, H.D. (1973), *J. Zool.*, Lond., **171**, 489–498.
[8] Lighthill, M.J. (1969), *Ann. Rev. Fluid Mech.*, **1**, 413–46.
[9] Parrott, G.C. (1970), *J. exp. Biol.*, **53**, 363–374.
[10] Pennycuick, C.J. (1968), *J. exp. Biol.*, **49**, 509–26.
[11] Pennycuick, C.J. (1971), *J. exp. Biol.*, **55**, 833–845.
[12] Pennycuick, C.J. (1972), *Animal Flight*, Arnold, London.
[13] Pugh, L.G.C.E. (1971), *J. Physiol.,* Lond., **213**, 255–276.
[14] Taylor, C.R. (1973), In Bolis, L., Schmidt-Nielsen, K. and Maddrell, S.H.P. (eds.) *Comparative Physiology*, pp. 23–42. North-Holland, Amsterdam.
[15] Taylor, C.R., Dmiel, R., Fedak, M. and Schmidt-Nielsen, K. (1971), *Am. J. Physiol.*, **221**, 597–601.
[16] Taylor, C.R., Schmidt-Nielsen, K. and

Raab, J.L. (1970), *Am. J. Physiol.*, **219**, 1104–1107.

[17] Thys, H., Faraggiana, T. and Margaria, R. (1972), *J. appl. Physiol.*, **32**, 491–4.

[18] Trueman, E.R. (1968), *Symp. zool. Soc. Lond.*, **22**, 167–186.

[19] Tucker, V.A. (1968), *J. exp. Biol.*, **48**, 67–87.

[20] Tucker, V.A. (1972), *Am. J. Physiol.*, **222**, 237–245.

[21] Tucker, V.A. (1970), *Comp. Biochem. Physiol.*, **34**, 841–846.

[22] Webb, P.W. (1971), *J. exp. Biol.*, **55**, 521–540.

[23] Weis-Fogh, T. (1972), *J. exp. Biol.*, **56**, 79–104.

[24] Weis-Fogh, T. (1973), *J. exp. Biol.*, **59**, 169–230.

5 Human mechanics

This chapter is about some aspects of human mechanics which have not been dealt with in earlier chapters. Many of the biological materials discussed in Chapter 2 are found in man, and much of the information about their properties was collected by scientists who felt it was likely to be useful to orthopaedic surgeons. The mechanics of human locomotion was considered in Chapter 4, together with the locomotion of other animals. The investigations described in this chapter seem to have been undertaken mainly to get a better understanding of the human body, though many of the experiments were performed on other mammals. They concern the lubrication of joints, the flow of blood and the inflation of lungs, all topics of considerable medical interest. There is an important related branch of research concerned with the design of replacements for defective parts of the body, such as joints and heart valves, which is not dealt with in this book.

5.1 Lubrication of joints

Osteoarthritis is a common and troublesome complaint. Joint surfaces become worn and joints become stiff and painful. A lot of research on normal joints has been done with the aim of discovering what goes wrong in osteoarthritis.

A well-lubricated joint has a low coefficient of friction. The coefficients of friction of mammal joints have been measured in various ways, for instance by using the joint as the pivot of a pendulum and observing how fast oscillations of the pendulum decay. Coefficients between about 0·002 and 0·01 have been found. This is the usual range for well-lubricated engineering joints. Such low coefficients of friction are apparently only possible when a film of fluid separates the solid surfaces. In engineering bearings the fluid is usually oil. In mammal joints it is synovial fluid, which resembles blood plasma but contains a proportion of a polysaccharide called hyaluronic acid.

When a stationary shaft rests on its bearings the oil is squeezed out from under it, but if the shaft is rotating it drags oil round with it so that there is always a film of oil between the shaft and the bearing. This effect plays a very important role in the lubrication of machinery but cannot be very important in mammal joints because they have a reciprocating (backward and forwards) rather than a rotary action, and because they often move slowly and frequently come to rest. There must be a different mechanism to keep the synovial fluid between the joint surfaces.

The synovial joints of mammals are enclosed in capsules of connective tissue, filled with synovial fluid (Fig. 5.1). The articulating surfaces of the bone are covered by a layer of cartilage, which is a few millimetres thick in man. The surface of the cartilage is remarkably rough. The roughness of engineering surfaces is measured by a device like a gramophone stylus which is drawn across the surface. The same technique has been used on plastic casts of articular cartilage surfaces (the cartilage itself is too

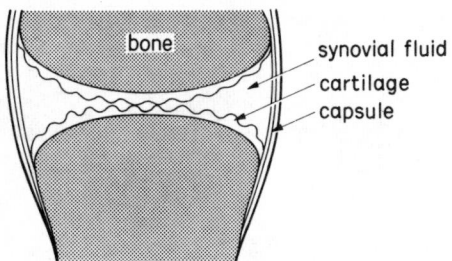

Fig. 5.1 A diagrammatic section through a typical mammalian joint.

soft). It is found that cartilage surfaces have irregularities which are about 2·5μm deep, peak to trough, in young, healthy joints and much more in old, arthritic ones. Even the healthy joints are much rougher than engineering bearings. When the cartilage is examined with a scanning electron microscope it can be seen that the irregularities are dimples bounded by bundles of collagen fibres [11]. The dimples are greatly exaggerated in Fig. 5.1.

The dimples probably play an important part in lubrication. Where the cartilage surfaces are in contact, synovial fluid is trapped in the dimples and cannot easily be squeezed out. The basic principle was explained by Charles McCutchen who named it 'weeping lubrication' and demonstrated it by experiments with foam rubber [7]. The rubber he used was the sort which has individual bubbles of air in it, not the sort which has a network of pores like a sponge. He cut it to open up a layer of bubbles, producing a deeply dimpled surface. He than rubbed it on glass with soapy water as a lubricant. The coefficient of friction was only 0·003. The water was trapped in the cut bubbles. Since the rubber between the bubbles was very easily deformable it carried little of the load, but it prevented the water (which bore most of the load) from being squeezed out sideways. The same sort of thing presumably happens in synovial joints.

The theory was taken a stage further by Duncan Dowson and his colleagues at Leeds [11]. They pointed out that cartilage has pores in it which are big enough for water to be squeezed through, but too small for the large molecules of protein and hyaluronic acid. When a joint is bearing a load, water and salts from the synovial fluid trapped in the dimples must be squeezed gradually out through the pores in the cartilage. They must leave behind the protein and hyaluronic acid, which become more and more concentrated. This phenomenon is called 'boosted lubrication'.

In addition to these effects, the elastic deformability of the cartilage may have a beneficial effect on lubrication: there may be elastohydrodynamic effects like those which play a major part in the lubrication of gear wheels.

5.2 Blood

Blood poses a remarkable range of problems in biomechanics, and a great deal of skill and effort has been devoted to solving them [5].

Human red corpuscles are discs of diameter 8μm, but some of the capillaries they flow through have diameters as small as 5μm. The corpuscles are squashed out of shape as they travel along these capillaries. They travel in single file, separated by plasma.

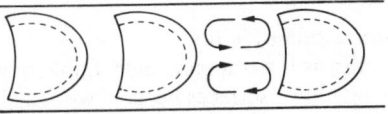

Fig. 5.2 A diagram representing red blood corpuscles moving from left to right in a narrow capillary. Movement of the plasma relative to the corpuscles is indicated by arrows.

Fluid in an unblocked tube flows fastest in the centre and more slowly near the walls. This is because fluid near the walls is slowed down by its own viscosity. In a capillary the tight-fitting corpuscles break the plasma into short sections. The plasma near the walls is slowed down by viscosity and so travels more slowly than the corpuscles. The plasma in the centre of the capillary compensates for this by travelling faster than the corpuscles, so the plasma circulates as shown in Fig. 5.2 as it travels along the capillary. (Water trapped between air bubbles moving along a tube circulates in the same way.) This circulation of the plasma must facilitate diffusion between the blood and the tissues. For instance, oxygen diffusing out of the centre of a corpuscle will not have to diffuse to the capillary wall, but will be carried round to the wall in the circulating plasma.

The movement of the tight-fitting corpuscles in the capillaries also poses a problem in lubrication [4]. A thin film of plasma can be expected to form between the corpuscle and the capillary wall, providing lubrication. The thickness of the film depends on elastic distortion of the corpuscle, so the problem is one of elastohydrodynamic lubrication. It is complicated by the capillary wall being porous, so that plasma can leak out into the surrounding tissue. Mathematical analysis indicates that changes in the pressure gradient along a capillary are likely to have disproportionately large effects on the velocity of flow and that the lubricating film will become very thin at low flow rates. This suggests that incomplete constriction of arterioles may slow down flow enough to make lubrication break down, so that flow stops althogether although the arterioles are not actually blocked.

In vessels much larger in diameter than the corpuscles, blood can be expected to flow as if it were a pure liquid. Poiseuille's equation describes steady, laminar flow of fluids along cylindrical tubes. It does not apply to pulsating flow such as occurs in arteries, nor does it apply to the turbulent flow which occurs when fluids flow fast or in wide tubes (i.e. in circumstances where the Reynolds number is high). Turbulent flow probably does not occur in any blood vessels except, perhaps, some large arteries. The equation gives the pressure difference P needed to drive a fluid of viscosity η at mean velocity $\bar{u}$ along a tube of length l and radius r. Though $\bar{u}$ is the mean velocity, the velocity varies from zero at the walls of the tube to $2\bar{u}$ in the centre. The equation is

$$P = 8\eta l\bar{u}/r^2 \qquad (5.1)$$

The viscosities of liquids are often determined by measuring the rate at which they are driven along a tube by a given pressure difference. Poiseuille's equation can then be used to calculate the viscosity. When this is done with blood, curious results are obtained. [3], [5]. Blood plasma is found to be 1·2 times as viscous as water. Whole blood is more viscous, as would be expected because of the resistance of the blood corpuscles to deformation. However, the apparent viscosity of whole blood depends on the diameter of the tube it is flowing in. If the diameter is 1 mm or more, the viscosity calculated from Poiseuille's equation is about three times that of water. As the diameter is decreased below 1 mm the viscosity falls until it is only 1·6 times the viscosity of water at a diameter of 12μm. If the diameter is decreased still more, the corpuscles begin to fit the tube rather closely and the effect discussed in earlier paragraphs becomes important. It is not really appropriate to apply Poiseuille's equation to blood in tubes so fine that the corpuscles fit them tightly, but if the apparent viscosity is nevertheless calculated it is found to increase again as the diameter decreases below 12μm (Fig. 5.3).

The reason for the decrease in apparent viscosity, as the diameter decreases from 1 mm to 12μm, seems to be as follows. When blood flows in a tube the red corpuscles tend to avoid positions very close to the wall. This can be observed to happen: there tend to be no corpuscles within about 4 μm of the wall. This can be

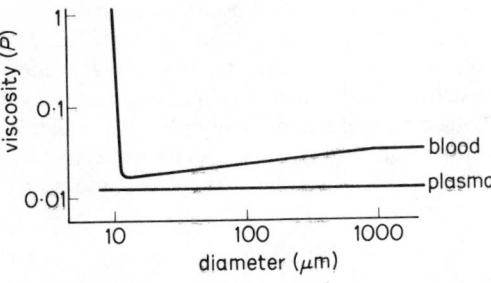

Fig. 5.3 A graph of the apparent viscosity of blood, against the diameter of the tube in which it is flowing. The viscosity of water at 20°C is 0·010 poise. Based on data in Refs. [3], [5],

explained by (rather elaborate) hydrodynamic arguments. If there are no corpuscles in the slow-moving plasma near the walls, the average velocity of the corpuscles is greater than the average velocity of the plasma. A corpuscle remains in the tube for a shorter time than a molecule of plasma. Hence the mean concentration of corpuscles in the tube is reduced, and with it the viscosity.

Veins have relatively thin, flexible walls and the pressure of the blood in them is low. They may get flattened by pressure from surrounding tissues. The effect of this on flow through veins has been studied by experiments with veins and with rubber tubes, and by computer simulation [6], [9]. The investigations show that in certain circumstances veins may oscillate, opening and closing in rapid succession.

Arteries present even more complex problems in mechanics. Their walls are too thick for them to be easily flattened, but they are elastic. Increasing the pressure in an artery makes it swell, so the arteries swell and contract again at each heart beat. The proximal part of the aorta of the dog, for instance, swells to about 1·1 times its minimum diameter [8].

The simple model shown in Fig. 5.4 may give some insight into the mechanics of flow in the arterial system. The heart is represented by a reservoir of liquid and the arterial system by a

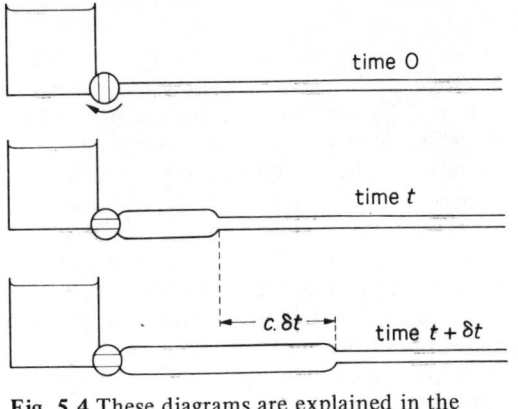

Fig. 5.4 These diagrams are explained in the text.

long uniform elastic tube. They are connected by a tap which is opened to represent systole and closed to represent diastole. The pressure is p at the distal end of the tube and $(p + \Delta p)$ in the reservoir. The liquid which represents the blood has density ρ and negligible viscosity.

Initially the tap is closed, the pressure all along the tube is p and the cross-sectional area of the tube is A. The tap is then opened and liquid flows out of the reservoir into the near end of the tube, distending it until the pressure in it is $(p + \Delta p)$ and its cross-sectional area is $A(1 + k.\Delta p)$ (k is a term representing the elastic compliance of the tube). The pressure further along the tube does not rise immediately: it cannot rise until more liquid flows into that part of the tube and distends it. Consequently a wave of distension travels along the tube.

Let the wave travel along the tube with velocity c. Since the cross-sectional area increases by $Ak.\Delta p$ as the wave reaches it the rate at which the volume of the tube increases is $Akc.\Delta p$. The rate at which blood enters the tube is $A(1 + k.\Delta p)u$, where u is the velocity of the blood. Therefore

$$Akc.\Delta p = A(1 + k.\Delta p)u$$

$$(1 + k.\Delta p)u/\Delta p = kc \qquad (5.2)$$

The liquid is still stationary in the undistended part of the tube but it is flowing with velocity u in the distended part. The rate at which the volume of moving liquid is increasing is $A(1 + k.\Delta p)c$ and the rate at which the momentum of the liquid is increasing is $\rho A(1 + k.\Delta p)cu$. This must be equal to the force $A.\Delta p$ which is acting on the liquid. Hence

$$\rho A(1 + k.\Delta p)cu = A.\Delta p$$

$$(1 + k.\Delta p)u/\Delta p = 1/\rho c \qquad (5.3)$$

From (5.2) and (5.3)

$$kc = 1/\rho c$$

$$c = (1/\rho k)^{\frac{1}{2}} \qquad (5.4)$$

This equation shows that the more extensible the tube is (i.e. the bigger k is) the slower the wave of distension will travel. Measurements on dogs show that the wave of distension (the pulse wave) travels along the proximal part of the aorta at 5 m s^{-1} and along the femoral artery (which has a less extensible wall) at 10 m s^{-1} [8].

This is a simple treatment of a complex problem. A real heart is not just a reservoir with a tap which is opened and closed instantaneously: the pressure in the ventricle rises to a peak and falls again. The walls of real arteries are not perfectly elastic, but visco-elastic, and real blood has viscosity. The arterial system is not an infinite uniform tube but branches repeatedly and waves of distension are liable to be reflected back from the branching points. All these points have to be considered if the form and velocity of the pulse wave are to be understood [5].

5.3 Surface tension in the lungs

The respiratory distress syndrome is a dangerous disease of newborn, and particularly premature, babies. It has been estimated that it accounts for 40% of deaths of premature babies. The baby breathes abnormally fast until it becomes exhausted, and often dies in a few days. Post-mortem examination shows that most of the alveoli of the lungs are collapsed, with no air in them. The lungs have the consistency of liver [2]. What has gone wrong?

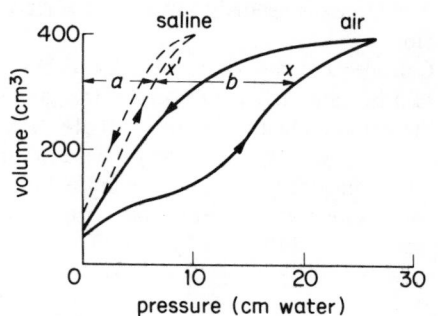

Fig. 5.5 Graphs of volume against pressure for dog lungs filled with air (continuous line) and saline (broken line). Redrawn from Ref. [1].

To understand the disease one must understand the mechanics of normal lungs, and the part played by surface tension. Fig. 5.5 shows typical results from an experiment which has been performed many times, with various modifications, over the past 45 years. A mammal's lungs are inflated with air and allowed to contract again, and a graph is drawn of their volume against the pressure inside them. They are then filled with physiological saline solution and the experiment is repeated. It is found that much smaller pressures are needed to inflate them when they are filled with saline, than when they are filled with air. It is also found, incidentally, that the graphs for inflation and deflation are not identical, but form loops.

The inner surface of the air-filled lung is moist and surface tension acts at the air/fluid interface. When the lung is filled with a saline

43

solution resembling the fluid of the lung surface, surface tension no longer acts. This is why the graphs for air-filled and saline-filled lungs are not identical. Consider the point X on the graph for the air-filled lung and the corresponding point X' for the saline-filled lung (Fig. 5.5). The total pressure acting at X consists of a component a attributable to tissue elasticity and a component b attributable to surface tension. It appears that most of the pressure needed to inflate the lungs is needed to overcome surface tension.

Consider a bubble of gas of radius r, immersed in a liquid. The pressure of the gas and of the liquid immediately around it are p_1, p_2, respectively. p_1 is greater than p_2 because surface tension·adds to the pressure in the bubble. It can be shown that if the surface tension in the gas/liquid interface is T,

$$p_1 - p_2 = 2T/r \qquad (5.5)$$

The surface tension in air/water interfaces is about $0.07 \, \mathrm{N\,m^{-1}}$. The value of b when the lung is fully inflated (Fig. 5.5) is about $1500 \, \mathrm{N\,m^{-2}}$ and by equation (5.5) this is the value of $(p_1 - p_2)$ for air bubbles of radius $2 \times 0.07/1500 = 10^{-4} \, \mathrm{m} = 0.1 \, \mathrm{mm}$, in water.

The air passages in the lung branch repeatedly and finally end in the little pockets known as alveoli. These are not complete bubbles but surface tension in the fluid film on their walls must produce the same pressure as if they were bubbles of the same radius. Typical alveoli in the inflated dog lung would have radii of the order of $0.1 \, \mathrm{mm}$ so the value of b is about what would be expected if the alveoli were moistened with pure water. However, the value of b for the deflated lung seems far too small. Equation (5.5) shows that the pressure difference due to surface tension should increase as the radius diminishes. Fig. 5.5 shows that b *decreases* as the lungs are deflated and the radii of the alveoli diminish. The data shown in the graph cannot be explained by any constant

value of surface tension. It can however be explained by the properties of a surfactant present in the lung.

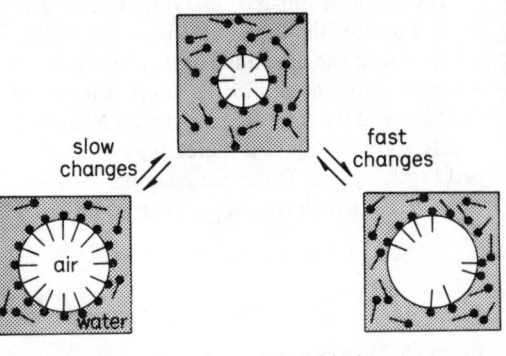

Fig. 5.6 Diagrams of an air bubble in an aqueous solution of a surfactant. The surfactant molecules are represented by lollipop symbols with the head at hydrophilic end.

Surfactants are soaps, detergents and other compounds with similar properties. They are more or less soluble in water because one end of the molecule is hydrophilic, but the other end is hydrophobic. Surfactant molecules in water tend to congregate at the surface. They form a single layer covering the surface with the hydrophobic ends protruding, as shown at the surface of the bubble in Fig. 5.6. Any excess molecules remain in solution. The surface film reduces the surface tension.

Fig. 5.6 shows what happens if the bubble is made to expand and contract repeatedly (presumably by changes of pressure). If the volume changes slowly molecules are forced out of the surface film during contraction and taken up again during expansion. The surface film is always complete and the surface tension is always low. However, if the volume changes rapidly the molecules cannot move in and out of the film fast enough. The surface film is incomplete when the bubble is expanded (Fig. 5.6, right). The surface tension is low when the bubble is small, but higher when it is large and

44

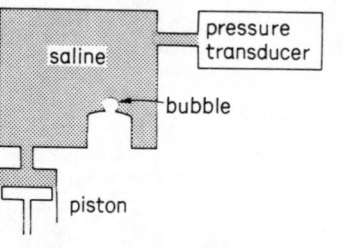

Fig. 5.7 Apparatus for investigating the properties of lung surfactant.

the film is incomplete.

There is a surfactant in mammal lungs which can be extracted by washing out with saline solution. A lipoprotein seems to be the active ingredient. Fig. 5.7 shows one of the methods which have been used to investigate its properties [10]. The chamber is filled with saline solution but a tiny bubble of air protrudes into it through a hole. The piston moves rhythmically up and down making the bubble contract and expand. The pressure transducer measures the pressures required. An image of the bubble is projected onto a photoelectric cell which is used to record its changes in radius. The surface tension at any instant can be calculated from the pressure and the radius. Surfactant is introduced into the solution as required.

Experiments with this equipment showed as expected that the surface tension has a minimum value when the oscillating bubble is contracted, and a maximum value when it is expanded. The maximum and minimum values change a little from cycle to cycle at first, but eventually a steady state is reached. The minimum found in experiments with dog lung surfactant was always about $0.02 \, \mathrm{N\,m^{-1}}$. The maximum depended on the amplitude and frequency of the oscillations but was about $0.07 \, \mathrm{N\,m^{-1}}$ (or about the same as pure water) when the oscillations were large and the frequency reasonably high (0.3 Hz). This change in surface tension is not dramatic enough to explain Fig. 5.5 quantitatively, but lower minimum surface tensions have been obtained by other techniques. Lower minimums (of the order of $0.001 \, \mathrm{N\,m^{-1}}$) were also obtained by the method shown in Fig. 5.7 in experiments with synthetic dipalmitoyl-lecithin. This compound is a constituent of lung surfactant extracts and it is not clear why it is more effective than the extract.

The repiratory distress syndrome can now be explained. The lungs of babies suffering from it are found to have abnormally little surfactant. The surfactant normally appears late in gestation, so premature babies are apt to have too little of it. If so, the surface tension in their lungs is high and does not diminish much as the alveoli contract. As an alveolus contracts with the surface tension remaining high, the pressure needed to keep it open increases. It is apt to collapse completely, and if it does collapse a large pressure will be needed to re-inflate it.

References

[1] Bachofen, H., Hildebrandt, J. and BA Bachofen, M. (1970), *J. appl. Physiol.* **29**, 422–31.

[2] Cherniack, R.M., Cherniack, L. and Naimark, A. (1972), *Respiration in Health and Disease*, ed. 2. Saunders, Philadelphia.

[3] Dintenfass, L. (1967), *Nature, Lond.*, **215**, 1099–1100.

[4] Fitz-Gerald, J.M. (1969), *Proc. R. Soc.* B, **174**, 193–227.

[5] Fung, Y.C. (1971), *Adv. appl. Mech.*, **11**, 65–131.

[6] Katz, A.I., Chen, Y. and Moreno, A.H. (1969), *Biophys. J.*, **9**, 1261–79.

[7] Lewis, P.R. and McCutchen C.W. (1959), *Nature, Lond.*, **184**, 1285.

[8] McDonald, D.A. (1960), *Blood flow in arteries*, Edward Arnold, London.

[9] Moreno, A.H., Katz, A.I., Gold, L.D. and Reddy, R.V. (1970), *Circulation Res.*, **27**, 1069–80.

[10] Slama, H., Schoedel, W. and Hansen, E. (1973), *Respiration Physiol.* **19**, 233–43.

[11] Wright, V., Dowson, D. and Kerr, J. (1973), *Int. Rev. Connective Tissue Res.*, **6**, 105–23.

6 Plant mechanics

The physical properties of wood and cellulose were considered, with those of other biological materials, in Chapter 2. Stresses in a branch were considered briefly in Chapter 3. This chapter is about some other topics in the mechanics of plants: the ascent of sap, the opening of stomata and the aerodynamics of winged seeds.

6.1 How sap rises

How does sap get to the tops of tall trees? There are two ways of getting water up a vertical pipe: it can be pumped up from below, or it can be sucked up from above. In the first method, a high pressure is applied at the bottom of the pipe. In the second, a low pressure is applied to the top. In either case there is a gradient of pressure up the pipe. If the water were stationary, the pressure would fall by 1 atm for every 10 m rise in height. If the water is flowing the gradient must be steeper, to overcome the viscosity of the water.

Ordinary suction pumps can only raise water rather less than 10 m. The reason is that once the pressure falls below the vapour pressure of water (0·02 atm at 20°C), bubbles of water vapour tend to form, or at least existing bubbles tend to grow. It is much harder to start a new bubble than to enlarge an existing one because surface tension makes the pressure in the bubble higher than the pressure in the surrounding water, and the effect is greatest if the bubble is small. The extra pressure produced by surface tension is $2T/r$ where T is the surface tension ($0·073 \mathrm{~N m}^{-1}$ for water) and r is the radius of the bubble. The molecules in water are about 0·3 nm apart so when a bubble first forms as a gap between molecules it must have a radius of the order of 0·15 nm. The pressure produced by surface tension in a bubble of this size is $2 \times 0·073/0·15 \times 10^{-9} = {}^{(}10^9 \mathrm{~N m}^{-2}$ or 10^4 atm. If the pressure *inside* a bubble of this size is to be below the vapour pressure of water the pressure *outside* must be 10^4 atm below the vapour pressure of water. If this were the whole story, a pressure of -10^4 atm would be needed to form new bubbles in bubble-free water. This can be expressed differently by saying that the theoretical tensile strength of water is about 10^4 atm or $1 \mathrm{~GN m}^{-2}$.

Various experiments have been performed which confirm that in favourable conditions water has a high tensile strength, though the measured strengths are far below the theoretical strength. For instance, it has been shown by spinning capillary tubes of bubble-free water in a centrifuge that pressures of -260 atm can be reached without bubbles forming.

The tallest trees are specimens of *Sequoia* and *Eucalyptus* about 100 m high. The pressure difference between the sap at the top and the sap at the bottom must be at least 10 atm. If the sap is pumped up from below the pressure at the bottom must be at least 11 atm (atmospheric pressure + 10 atm). If it is sucked from above, the pressure at the top must be more negative than -9 atm (10 atm below atmospheric pressure). Knowledge of the tensile strength of water suggests the possibility that the trees may do what ordinary pumps could not, and draw water up their great height from above.

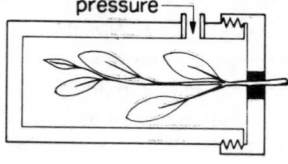

pressure

Fig. 6.1 A device for determining the pressure in xylem sap.

Columns of water at negative pressure are unstable. Any disturbance is liable to enable bubbles to form, so it is no use looking for negative pressures simply by attaching manometers to trees. An experiment which will do the job, albeit indirectly, was devised by P.F. Scholander and his colleagues [5]. Consider a twig on a tree with its sap at a pressure of $(1 - X)$ atm. There is a pressure difference of X atm between the sap and the exterior, so the tissues are under stress. If the twig is cut the sap pressure is allowed to rise to 1 atm and the pressure difference disappears. The tissues which have been under stress expand, drawing sap up the stem away from the cut. The effect is accentuated because the contents of the living cells in the twig have a higher osmotic pressure than the sap and are enabled by the change in sap pressure to take up water osmotically from the sap. These changes can be reversed in the 'bomb' shown in Fig. 6.1. It is a vessel made strong enough to withstand high pressures. The twig is fitted into it as shown and the pressure is increased until the sap is forced back to the cut surface and just begins to seep out. This will happen when the pressure in the bomb is $(X + 1)$ atm. The xylem is open to the atmosphere at the cut surface so the pressure in the xylem is 1 atm and the pressure difference between the sap and the air surrounding the leaves is X atm, as it was when the twig was on the tree.

It is not easy to cut twigs from the tops of really tall trees so a marksman was employed to shoot twigs down with a rifle (this is not easy either). In a typical experiment twigs were shot off a redwood (*Sequoia*) at heights of 15 and 82 m. It was found that the sap pressures were −8 atm and −15 atm, respectively. The difference of 7 atm for almost 70 m represents a gradient of about 0·1 atm m^{-1}. This is about the minimum gradient needed to overcome gravity, but the pressure must have been strongly negative even in the roots.

Negative pressures may be needed in roots, to withdraw water from dry soil where it is held by capillarity in the spaces between soil particles or absorbed by colloidal materials. Strongly negative pressures may be found in smaller plants in dry places, as well as in tall trees. Twigs taken from a creosote bush after many months of drought were tested by the bomb technique. The sap pressure was found to be −80 atm. Strongly negative pressures are also found in mangroves, which get their water from the sea although their sap is essentially fresh water. The sea and the sap are separated by a layer of cells in the roots which acts as a semipermeable membrane. The water from the sap would diffuse out into the sea if it were not held back by a sufficiently large pressure difference. The osmotic pressure of sea water is about 25 atm so the hydrostatic pressure of the sap must be −24 atm (24 atm below atmospheric), or more negative, if water is to be withdrawn from the sea. Pressure of −40 atm, and even more negative, have been recorded by the bomb technique.

It has been argued recently that the bomb experiments have not after all demonstrated negative pressures in the sap [3]. Suppose the xylem vessels were not continuous tubes, but consisted of short chambers set end to end and separated by semipermeable membranes. Suppose they were filled with solutions of different concentrations, increasing from the bottm to the top of the tree. The gradient of osmotic pressure could counteract the effect of gravity. Water could be drawn up the tree without there being any gradient of hydrostatic pressure. Of course the vessels are not like this, but a similar effect could be obtained in a different way. The vessels could contain a gel of polymer strands, firmly anchored to their walls and increasing in concentration from the bottom to the top of the tree. No semipermeable membranes would be needed to separate the different concentrations, because the gel was anchored. The idea is interesting, but evidence for the existence of the gel seems tenuous.

The gradient of hydrostatic pressure (or of gel concentration) is needed to overcome the viscosity of the water as it is driven through the narrow tubes of the xylem, as well as to overcome gravity. How much does this add to the pressure gradient? In some trees such as the redwoods it apparently adds very little, for the gradients found by the bomb technique are barely more than the minimum of 0.1 atm m^{-1} needed to overcome gravity. In other trees it may add considerably more.

The additional gradient must depend on the speed of the sap and the dimensions of the vessels. The speed can be measured in various ways. For instance, a small electric heater can be embedded under the bark, and a thermocouple embedded a little further up the same tree. The heater is switched on briefly and a little later the thermocouple registers an increase in temperature. The interval is the time taken by the sap which was heated to rise to the thermocouple, so the rate of ascent can be calculated. In some species, very high rates are found. For instance, it has been found that sap may rise as fast as 44 m hr^{-1} in oak trees (*Quercus*). In other species, the rates are much lower. The pressure gradient needed to drive the sap at the measured speeds can be determined by measuring the pressures needed to force water along cut lengths of wood. The data

which is available is rather patchy, but seems to show that pressure gradients of 0.05 to 0.1 atm m^{-1} are likely to be needed at the times of day when sap flows fastest, in addition to the 0.1 atm m^{-1} needed to counteract gravity [9].

The pressure gradient needed for a given rate of flow can also be calculated from the diameter of the xylem vessels and the viscosity of water using Poiseuille's equation (see p. 42). The calculated gradients are generally much lower than those measured in experiments with cut lengths of wood. The explanation is that the xylem does not consist of simple uninterrupted tubes. The vessels of angiosperms consist of short segments, much as a sewer is built of short lengths of pipe (Fig. 6.2a). There are no partitions across a vessel, but no vessel is as tall as the tree. It requires some ingenuity to find out just how long the vessels are, but methods have been devised. It has been found for instance that *Eucalyptus* vessels are up to 3 m long, averaging about 0.5 m [9]. Water can flow freely along a vessel but when it comes to the end it must pass through small pores in the pits to another vessel. It is because it has to be forced through these pores that more pressure is needed than Poiseuille's equation (assuming continuous pipes) predicts.

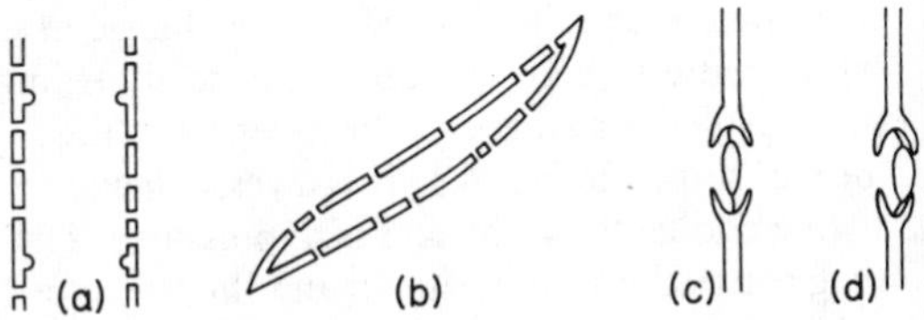

Fig. 6.2 Sections of (a) a vessel, (b) a tracheid, (c) a bordered pit and (d) the same, with the central disc displaced by pressure from the left.

Conifers do not have vessels, but tracheids. These are only $0.5 - 3$ mm long (Fig. 6.2b). The sap has to pass through pits to get from one tracheid to the next, and the pits must have a large effect on the pressure needed to drive it.

The mechanics of the effect has not been fully elucidated [4], [8].

Sap would flow more easily up trees if its path were not interrupted by pits, but the interruptions may serve an important function. Since columns of water under tension are inherently unstable, the columns of sap in the xylem must sometimes break, forming bubbles of vapour. It will be shown that the pits will tend to confine the damage to a single vessel or tracheid. Breaking of sap columns can be demonstrated. Turgid leaves have been cut, and the cut surface of the petiole sealed with wax. The petiole has then been impaled on a needle mounted on a gramophone pick-up. As the leaf dries out the sap in vessels breaks and clicks are recorded [9].

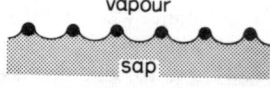

Fig. 6.3 This diagram is explained in the text.

Consider a vessel of tracheid filled with sap at a negative pressure. It the sap breaks, the bubble of vapour which is formed will tend to grow and fill the whole vessel, so that sap can no longer flow along it. However, the bubble cannot easily spread to other vessels. This is because there are webs of fine filaments forming membranes across the pits. The membranes are porous so sap can flow through them, but the pores are very small. For instance, electron micrographs of pit membranes of lime (*Tilia*) show pores of radius about 20 nm [9]. As a bubble in a vessel grows and the sap retreats through a pit, the sap surface will form menisci across the pores (Fig. 6.3). Surface tension in a hemispherical meniscus can withstand a pressure difference $2T/r$, where T is the surface tension and r the radius. The surface tension of water is 0·073 $N\,m^{-1}$ so pores of radius 20 nm (2×10^{-8} m) should be able to hold back a bubble against a pressure difference of $2 \times 0·073/2 \times 10^{-8} = 7 \times 10^6\ N\,m^{-2} = 70$ atm. The pits should be able to stop bubbles from spreading even when the pressure in the next vessel is exceedingly low. The mesh of the web in conifers is much less fine, but a solid disc is suspended in the centre of the membrane (Fig. 6.2c). Large pressure differences must tend to force the disc out of position so that it blocks the pore (Fig. 6.2d). This valve action should help to prevent bubbles from spreading.

6.2 How stomata open

Leaves are covered by an epidermis with a more or less impermeable cuticle, which is perforated by stomata. The stomata open and close, and thus control to some extent the rate at which water is lost by evaporation. The conditions in which stomata open are discussed in books on plant physiology. The mechanism of opening presents a problem in mechanics, which will be discussed here.

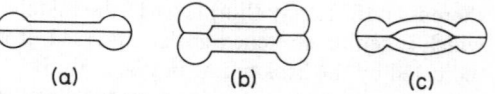

(a)   (b)   (c)

Fig 6.4 These diagrams of a grass stoma are explained in the text.

The edge of a stoma are formed by two guard cells. In grasses, the guard cells are shaped like dumbbells (Fig. 6.4a, which shows a closed stoma). In many other plants they are kidney-shaped, but only grass stomata will be considered here. There is plenty of evidence that the concentrations of dissolved sugars and ions in the guard cells increase as the stomata open. The guard cells must take up water osmotically. Swelling is resisted by their relatively inextensible cell walls, so the pressure inside them must rise. The rise in pressure presumably opens the stoma, but how does it do so? It is often said that the pressure makes the bulbous ends swell and so opens the stoma (Fig. 6.4b). However, this does not seem to be what actually happens. Comparison of photographs of the same stoma, open and closed, show that the ends swell very little. They certainly do not swell enough to account for the opening. The guard cells bow apart, as shown in Fig. 6.4c.

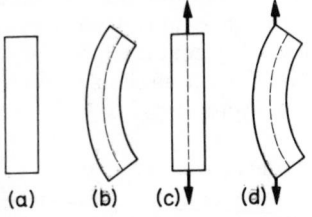

Fig. 6.5 These diagrams are explained in the text.

A possible mechanism has been found by careful study of the form of the cell wall in maize, *Zea* [7]. It depends on the theory of beams. Consider the block of elastic material shown in Fig. 6.5a. If it is bent one side of it gets stretched and the other compressed but there is an (infinitely thin) layer in the middle which is neither stretched nor compressed. It is indicated by the broken line in Fig. 6.5b. It passes through the centroid (geometrical centre) of each cross section. If forces pull on the block in line with the centroids it stretches without bending (Fig. 6.5c). If they pull to one side of the centroids, they cause bending (Fig. 6.5d).

The cell walls of the guard cells can be compared to this elastic block. They are not stretched by being pulled from outside, but by the pressure of the cell contents within them. The lengthwise component of the pressure can be considered to act through the centroids of the cross-sections of the cell interior. Two cross-sections are shown in Fig. 6.6. In each case the centroid of the cell interior has been located, by cutting it out from a photograph of the section and finding the centre of mass of the

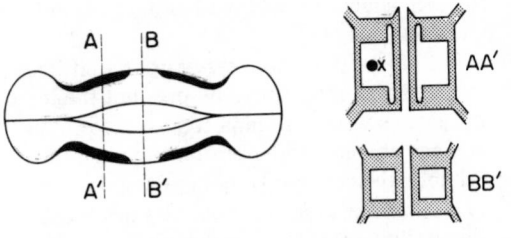

Fig 6.6 Diagrams of a stoma and of two sections through it, showing where the walls of the guard cells are thickened. The centroids of the cross sections of the cell walls (●) and of the cell contents (x) are shown in section AA'.

cutout. The centroids of the cell wall, as it appears in the sections, have been located in the same way. In section AA' the centroids of the interior and of the wall do not coincide. The pressure of the interior must tend to make the cells bend, concave outwards. In section BB' the centroids do coincide and there would be no tendency to bend there, were it not for the bending occurring on either side. Because of this bending the cells have to bend concave-inwards at BB', and form the bow shape characteristic of the open stoma. Calculations seem to confirm that reasonable pressures could probably open stomata in this way. However, the calculations are tentative because Young's modulus for lengthwise stretching of the cell walls is not known.

Only grass stomata have been discussed here, but an analysis of the mechanics of the thick-walled elliptical stomata possessed by most other terrestrial plants has been published [1].

6.3 Falling seeds
The seeds of sycamore and maple (*Acer*) and of some other trees spin in a pleasing way as they fall. Their spinning slows down their fall because they act like helicopter rotors. It makes them more likely to land well away from the parent tree in a favourable place for germination.

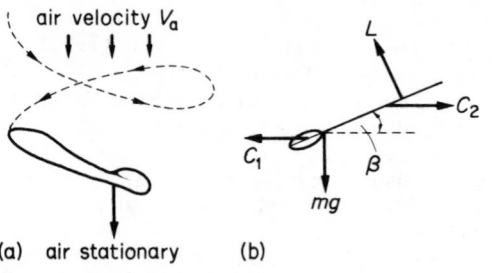

(a) air stationary   (b)

Fig. 6.7 Diagrams of a falling sycamore seed.

The rate of fall of an ideal sycamore seed can be calculated by means of the theory of propellers. Consider a seed falling with velocity V through still air (Fig. 6.7a). As it falls and rotates its tip describes a helix of radius r, cross-sectional area πr^2. The air below it is stationary but as it falls it imparts a downward velocity to the air it passes, so that a column of air above it has a downward velocity V_a. It can be shown that the velocity of the air as it passes the plane of the seed is $\frac{1}{2}V_a$. The velocity of the seed *relative to this air* is $(V - \frac{1}{2}V_a)$ so the volume of air which is passed by the seed in unit time and accelerated downwards is $\pi r^2(V - \frac{1}{2}V_a)$. If the density of the air is ρ the mass of air accelerated in unit time is $\rho\pi r^2(V - \frac{1}{2}V_a)$. This air eventually reaches a velocity V_a so the momentum given to the air in unit time is $\rho\pi r^2 V_a(V - \frac{1}{2}V_a)$. The force exerted by the air on the seed equals this rate of change of momentum. If the seed is falling with constant velocity the force is also equal to its weight, mg, and

$$mg = \rho\pi r^2 V_a(V - \tfrac{1}{2}V_a)$$
$$V = (mg/\rho\pi r^2 V_a) + \tfrac{1}{2}V_a .$$

The first term on the right hand side of the equation is large when V_a is small, and the second term is large when V_a is large. Obviously V has a minimum value at some intermediate value of V_a. It can be shown by differentiation

that the minimum value occurs when $V_a = 0.67V$ and is

$$V_{min} = 1.5(mg/\rho\pi r^2)^{\frac{1}{2}} .$$

This is only a little more than the rate of fall of a well-designed parachute of mass m and radius r. The parachute would be much bulkier than the winged seed, and it is hard to see how it could be made so light.

This is a theoretical minimum velocity. How nearly do real seeds approach the minimum? As an example, consider a maple seed of mass 0.13 g and spinning radius 3.5 cm. Its theoretical minimum sinking speed in air of density 1.3 kg m^{-3} is 0.75 m s^{-1}. Its actual sinking speed was 0.9 m s^{-1}, which is strikingly close to the minimum [2].

This simple application of propeller theory explains why such low sinking speeds are possible, but it leaves a lot unexplained. What starts the falling   l spinning in the first place? What keeps it spinning in stable fashion, correcting its attitude and angular velocity if they deviate from the optimum? What is the best shape for the blade? These questions have all been discussed in some detail [2], but only one aspect of stability will be considered here. The seed rotates with its blade at an angle β to the horizontal (Fig. 6.7b). How is this angle β kept constant at a suitable value? The seed rotates about a vertical axis which passes more or less through its centre of mass. Centrigugal forces C_1, C_2 exert moments on the seed tending to diminish β. The aerodynamic lift L, acting on the blade, exerts a contrary moment tending to increase β. At equilibrium, the moments are balanced. If β decreases the centrifugal forces come more nearly into line with the centre of mass and so exert smaller moments. The lift therefore pulls the blade back to the equilibrium angle. Similarly if β increases the centrifugal forces exert larger moments and reduce β again. The equilibrium is stable.

Thistledown and similar seeds do not spin, but have parachutes. They fall more slowly than maple seeds. For instance, the seed of *Cirsium arvense* falls at only 0.2 m s^{-1} [6]. This does not indicate that the parachute mechanism is superior, because we are comparing seeds of very different weight. Image a piece of giant thistledown, scaled up to have the same weight as a maple seed. If it was n times as long as ordinary thistledown it would be n^3 times as heavy but it would only have n^2 times the parachute area. It can be shown that it would fall $\sqrt{n}$ times as fast as ordinary thistledown: this might well be faster than a maple seed.

References

[1] De Michele, D.W. and Sharpe, P.J.H. (1973), *J. theoret. Biol.*, **41**, 77–96.

[2] Norberg, R.A. (1973), *Biol. Rev.*, **48**, 561–96.

[3] Plumb, R.C. and Bridgman. W.B. (1972), *Science, N.Y.*, **176**, 1129–31.

[4] Preston, R.D. (1974), *The Physical Biology of Plant Cell Walls*, Chapman and Hall, London.

[5] Scholander, P.F., Hammel, H.T., Bradstreet, E.D. and Hemmingsen, E.A. (1965), *Science, N.Y.*, **148**, 339–46.

[6] Sheldon, J.C. and Burrows, F.M. (1973), *New Phytologist*, **72**, 665–75.

[7] Shoemaker, E.M. and Srivastava, L.M. (1973), *J. theoret. Biol.*, **42**, 219–25.

[8] Smith, D.N.R. and Banks, W.B. (1971), *Proc. R.Soc.* B. **177**, 197–223.

[9] Zimmerman, M.H. and Brown, C.L. (1971), *Trees, Structure and Function*, Springer, Berlin.

7 Cell mechanics

This chapter describes some of the ways in which mechanics is being applied to the tiny structures which make up individual cells. There are sections about cilia and flagella which are used by some small organisms for swimming and by larger animals for moving fluids around the body; about cell division; about the myonemes which cause rapid changes of shape in ciliate protozoa; and about cell membranes.

7.1 Cilia and flagella

Cilia and flagella are hairlike, motile projections from cells [5]. They are identical in internal structure and there seems to be no really fundamental difference between them, but the two names are still used. The processes called flagella generally undulate like a swimming eel and drive fluid away from the cell surface (Fig. 7.1a). Those called cilia generally beat asymmetrically, driving fluid parallel to the cell surface (Fig. 7.1b). Flagella are generally found singly or in small numbers on cells but cilia occur in large numbers, packed closely together.

A spermatozoon swimming with its flagellum, suitably magnified, looks very like an eel swimming with its muscular tail. However, it presents a very different problem in hydrodynamics. A spermatozoon is very small and swims slowly. An eel is both much larger and much faster. One consequence of the difference is that the pattern of water flow around the spermatozoon is quite different from the pattern around the eel. The effects of the inertia of water can be neglected when the spermatozoon is being discussed, but are very important in the case of the eel. Hydrodynamic forces acting on the spermatozoon are proportional to its velocity relative to the water, but hydrodynamic forces on the eel are more or less proportional to $(velocity)^2$. Readers who are familiar with the concept of Reynolds number may like to know that the Reynolds number (based on body length) is of the order of 10^{-2} for a bull spermatozoon and 10^6 for a fairly large eel.

The forces which can be expected to act on undulating flagella can be calculated. It can be shown that the energy needed to propel a spermatozoon (or other flagellated organism) should be least if the amplitude of the waves is about 0·16 wavelength. This is the ratio of amplitude to wavelength shown in Fig. 7.1a, and flagella beating normally commonly show about the same ratio.

It is often assumed in analyses of the mechanics of flagella that their movements are sinusoidal. This is often mathematically convenient, but close study of photographs of flagella shows that their waves are not sine waves. They are more like a series of arcs of circles, joined by straight lines, as shown in Fig. 7.1a. The difference does not seem to make much difference to the forces produced [2].

The forces produced by cilia can be calculated in the same way as for flagella, but the calculation is complicated by the need to consider the effects of neighbouring cilia on each other. Solutions to some problems can be found in a different way. Neighbouring cilia beat slightly

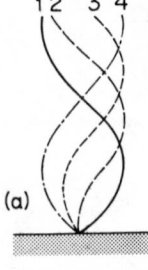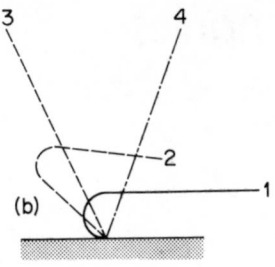

Fig. 7.1 Diagrams showing how typical flagella (a) and cilia (b) beat. Successive positions are shown by different styles of line, and numbered serially. The beat of many flagella and cilia is more or less confined to a plane but some flagella form helices rather than planar waves, and some cilia beat in distinctly three-dimensional fashion.

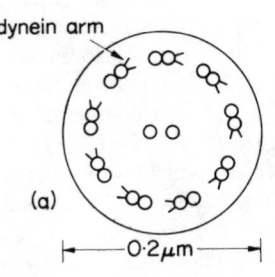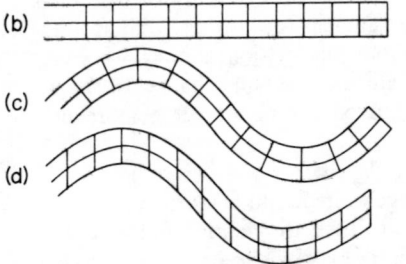

Fig. 7.2 (a) A transverse section through a cilium of flagellum. (b), (c) and (d) Diagrams of flagella which are explained in the text.

out of phase with each other so that ripples pass over a field of cilia like the ripples produced by wind in a field of corn. Some problems can be solved by treating the surface of a field of cilia as a continuous rippling membrane [1].

How do cilia and flagella work? What makes them bend? One possibility which has been considered is that they are passive projections, wagged by active structures in the body of the cell. Waves can be made to travel along a rope simply by shaking one end. However, if this was how flagella worked the waves would decay as they travelled along them, damped by the viscosity of the water. It has been calculated that the waves would have died out more or less completely by the time they had travelled $1\frac{1}{2}$ wavelengths along the flagellum. Flagella often

show more than $1\frac{1}{2}$ waves, and the amplitude of the waves may actually increase from base to tip. Hence flagella cannot be passive but must be capable of active bending.

Flagella and cilia are very uniform in internal structure. Sections examined in the electron microscope show a ring of fine double tubules surrounding two single ones (Fig. 7.2a). There is an outer membrane, and there are tenuous connections between the tubules. There are two ways in which bending could occur. Either the tubules on the inside of each bend could become shorter than the tubules on the outside (Fig. 7.2c) or the tubules could slide relative to each other (Fig. 7.2d). The latter seems to be what happens. Electron micrographs of complete, bent flagella show that the filaments on

54

the inside of the bend project beyond those on the outside, as at the right hand end of the flagellum shown in Fig. 7.2d. Notice how different this is from the arrangement of the end of the flagellum in Fig. 7.2c.

More dramatic evidence that sliding occurs has been obtained by experiments with flagella from sea-urchin sperm [8]. The flagella were separated from the cell bodies and then treated with trypsin, which partly disrupted their structure but left the tubules intact and undisturbed. When ATP was added to the treated flagella, they broke up into their component tubules. When the process was watched under the microscope, some of the flagella stuck to the coverslip. It could be seen that some of the tubules which had apparently not got stuck slid along the stuck ones until they came off at the ends and broke free. The ATP was making the tubules slide along each other. This would cause bending in an intact sperm, but it destroyed the treated ones.

In muscle, the myosin filaments have projections which attach to the actin filaments and apparently cause sliding. The double tubules of flagella have projections (dynein arms, Fig. 7.2a) which presumably have a similar function.

Fig. 7.3 shows some of the changes which occur as waves travel along a flagellum [3]. The graphs are drawn as they would be if the flagellum were infinitely long: more realistic assumptions about the length of the flagellum would make this explanation more complicated. As the waves travel along the flagellum each point on the flagellum moves from side to side. The transverse velocity is greatest between the bends in the flagellum, and falls to zero at each bend, as Fig. 7.3b shows. The fluid exerts forces on the moving parts of the flagellum, which are proportional to the velocity, and these forces exert bending moments on the flagellum. The bending moments are greatest, in this situation, where the force is greatest, just as the bending

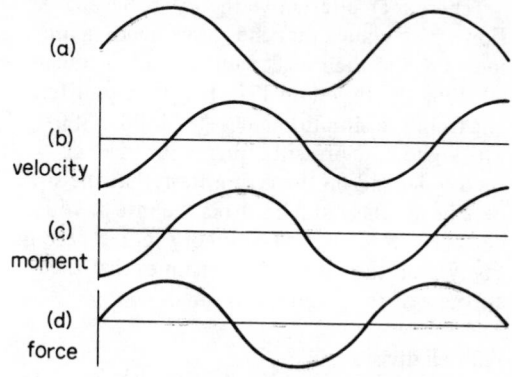

Fig. 7.3 (a) A diagram of a flagellum, with waves travelling from left to right. (b), (c) and (d) Graphs showing the distributions of velocity, bending moment and force per cross-bridge along the flagellum.

moment in a bridge has a maximum immediately under a heavy load. Hence the distribution of bending moments along the flagellum must be as shown in Fig. 7.3c.

The bending moments exerted by the fluid must be balanced by active bending moments produced in the flagellum by the forces in the cross-bridges. These forces must be greatest where the moments are changing most rapidly, that is where the moments are zero, at the bends in the flagellum (Fig. 7.3d). The required balance between the active forces and the fluid resistance will be achieved if the average force per cross-bridge at any point along the flagellum is roughly proportional to the curvature of the flagellum. If there is some mechanism which makes the cross-bridges exert forces proportional to the curvature, a wave started at the base of the flagellum will be propagated automatically along the flagellum without the need for any further mechanism of co-ordination. An eel needs a complex nervous system to co-ordinate waves of bending travelling along its body but flagella probably depend on an automatic response of the cross-bridges to bending.

There is an alternative theory of flagellar action which claims that the active bending moments rise to their peak values simultaneously all along the flagellum [7]. This is very different from the situation envisaged in Fig. 7.3. The arguments presented in favour of this theory depend on the assumption that the stiffness of the flagellum controls the rate at which bends travel along it. The theory represented in Fig. 7.3 on the other hand, assumes that the stiffness of the flagellum is small.

7.2 Cell division

What forces act when a cell divides? How is the cytoplasm separated into two halves? Many experiments have been performed to try to answer these questions, and sea-urchin eggs have often been chosen as the cells for study. A fertilized egg dividing in two is a simpler system to study than a tissue containing many dividing cells. Sea-urchin eggs are small and spherical, about 0·1 mm in diameter in some of the species which have been used for experiments. The outer (extracellular) membranes can be removed by treatment with urea, leaving the cell membrane exposed.

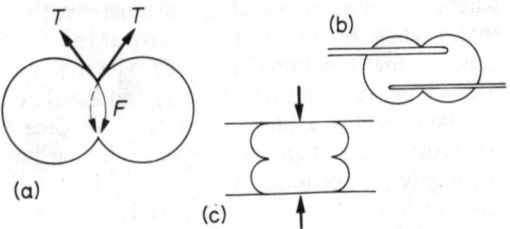

(a)

(b)

(c)

Fig. 7.4 (a) Forces acting at the 'waist' of a dividing cell. (b) and (c) Experiments on dividing cells which are described in the text.

The surface of an egg must be under tension, to maintain its spherical shape. When the egg divides (Fig. 7.4a) this tension, T, pulls outward on the 'waist'. A constricting force F is needed

to resist it. This force is presumably provided by a bundle of microfilaments which forms a ring encircling the waist. Microfilaments have about the same diameter as the thin filaments of muscle (p. 14) and seem to be chemically similar.

The constricting force has been measured in two ways, of which the more direct is shown in Fig. 7.4b. Two glass needles mounted on micromanipulators are pushed into the dividing egg, as shown. Both needles are necessarily thin but one is thinner (and so more flexible) than the other. As division proceeds the constricting ring presses on the needles, bending them. The amount the finer needle bends is observed, and used to calculate the force [6].

A less direct method is shown in Fig. 7.4c. The egg is compressed between two pieces of cover glass. The shape it adopts under a known compressing force can be used to calculate the tension, T. This in turn can be used to calculate the constricting force F needed for equilibrium. T and F both change in the course of division [11].

This method shows that F has a maximum value of about 6×10^{-3} dynes in *Pseudocentrotus depressus*. This occurs when the diameter of the waist is about 70% of the initial diameter. Forces of the same order of magnitude were measured by the direct method.

The bundle of microfilaments which is presumed to exert this force is about 6 μm wide and 0·1 μm thick so its cross-sectional area is about 6×10^{-13} m^2. If it exerts 6×10^{-3} dynes $= 6 \times 10^{-8}$ N the stress in it is 10^5 N m^{-2}, which is of the order of magnitude of the stresses exerted by striated muscle (p. 15).

Microfilaments seem to be responsible for many changes of cell shape, as well as the ones which occur in cell division [10]. However, there are other contractile organelles found in ciliate Protozoa. These are the myonemes, which are discussed in the next section.

7.3 Myonemes

The myonemes which have been studied most thoroughly are the ones called spasmonemes, in the stalks of peritrich ciliates such as *Vorticella* (Fig. 7.5a). It is believed that microfilaments work in much the same way as muscle, but it has become apparent that spasmonemes work on an entirely different principle [9].

Peritrich ciliates live in fresh water, attached to water plants by their stalks. The spasmoneme runs along the stalk and when it contracts it shortens the stalk by making it form a helix (Fig. 7.5a), or by bending it at a joint, or both, according to species. Shortening occurs very quickly and is presumably an escape reaction.

Muscle fibres treated with glycerol solutions can be made to contract by adding ATP. *Vorticella* stalks in glycerol can be made to contract in a quite different way, by adding calcium salts to the solution. In solutions containing 1 mEquiv Ca^{2+} l^{-1} the stalks contract, but when the concentration is reduced to 0·01 mEquiv Ca^{2+} l^{-1} they extend again. No ATP is needed and the process is not prevented by metabolic inhibitors. The process is obviously quite different from muscle contraction.

The spasmoneme is also quite different in structure from muscle. Electron microscope sections show that it consists of fine filaments, each like a chain of tiny beads, interspersed by tubules with membranous walls. The filaments are protein, and seem to be all alike. It seems likely that the tubules secrete and withdraw calcium, making the filaments contract and extend. ATP may be needed to provide the energy for moving the calcium, but any part it plays is obviously quite different from its part in causing muscle contraction.

The spasmoneme of *Vorticella* is so small that it does not seem practicable to measure its mechanical properties. Happily there is a colonial peritrich, *Zoothamnium,* which has a much larger spasmoneme. It consists of several thousand individuals with a common stem. They share a giant spasmoneme about 1 mm long. This is still too small to handle really easily, but its properties have been investigated in apparatus constructed on a microscope slide, very like the apparatus used to investigate the elastic tendon of dragonflies (p. 9).

Some of the results of the investigations are shown in Fig. 7.5b. The spasmoneme can be stretched greatly and will recoil elastically, both at low calcium concentrations and at high ones. It is shorter, at any given stress, at high concentrations than at low ones. When a spasmoneme is moved from a low to a high calcium concentration it shortens by almost a third if it is free to do so, or develops tension if its ends are held.

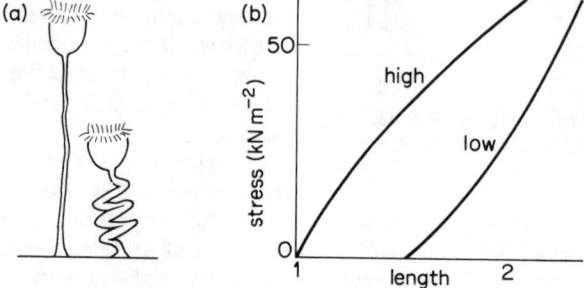

Fig. 7.5 (a) *Vorticella* with its stalk extended and contracted. (b) Graphs of stress against length for the spasmoneme of *Zoothamnium,* at high and low calcium concentrations. Lengths are expressed as multiples of the unstressed length at high calcium concentrations. Re-drawn from Ref. [9].

The results of these mechanical experiments and of measurements of birefringence indicate that the contracted spasmoneme is an amorphous cross-linked polymer like resilin (p. 9). Each bead in the protein filaments is a randomly coiled molecule. Removal of calcium allows additional cross-links to form and these hold the molecules in non-random, extended configurations.

Spasmonemes can shorten by 30% in only 3 ms. The rate of shortening is 100 lengths s^{-1}. This is faster than any known muscle. Even the extensor digitorum longus of mice, an exceptionally fast striated muscle, can only achieve 30 lengths s^{-1}. However, the spasmoneme takes several seconds to extend again and the stresses it can develop (around 50 $kN\,m^{-2}$, Fig. 7.4b) are low compared to the stresses developed by muscle. (Vertebrate striated muscle, contracting isometrically, develops about 300 $kN\,m^{-2}$, p. 15).

7.4 Cell membranes

Cell membranes consist of two layers of lipid sandwiched between layers of protein, each only one molecule thick (Fig. 7.6a). What are their mechanical properties?

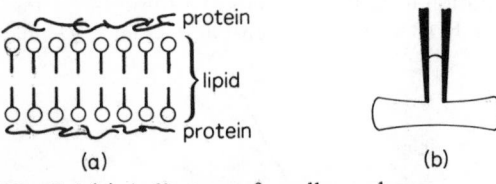

(a)   (b)

Fig. 7.6 (a) A diagram of a cell membrane. (b) An experiment described in the text.

Various experiments have been performed to investigate the properties of the outer membranes of red blood corpuscles [4]. Different experiments give different impressions. When a micropipette is applied to the surface of a corpuscle, it is easy to suck a nipple of

protoplasm up into the pipette (Fig. 7.6b). When the suction is released, the cell returns to its initial shape. This experiment gives the impression that the membrane is elastic and highly extensible. When a corpuscle is made to swell osmotically, it changes from a disc to a sphere and its volume increases greatly, but its surface area does not change appreciably. This experiment gives the impression that the membrane is highly inextensible. How can the paradox be resolved?

Compare the membrane to rubber. A rubber band is easily stretched lengthwise, and gets thinner as it is stretched so that its volume remains almost constant. Rubber in the wall of a balloon is easily increased in area, as the balloon is inflated. It gets thinner as its area increases, and its volume is hardly changed. It is easy to stretch rubber in length or in area, but it is very hard to change its volume; in technical terms, the bulk modulus of rubber is much higher than the shear modulus. Large changes in pressure have little effect on the volume of rubber.

If the area of a cell membrane were increased by stretching the lipid layers could not get thinner because they are already only one molecule thick. It is believed that each lipid layer behaves as a two-dimensional fluid. It probably offers no appreciable resistance to changes of shape of the membrane so long as the area remains unchanged, because the molecules can rearrange themselves to suit the change of shape. However, it resists changes of area. The protein layer is probably cross-linked, so that it has rubber-like elasticity and does resist changes of shape.

When a nipple of protoplasm is drawn up into a pipette (Fig. 7.6b), the area of the cell membrane is presumably unchanged but the network of protein molecules is distorted. The elasticity of the protein layer restores the shape of the cell when the suction is released.

The results of the pipette experiments, and of other experiments with red corpuscles, have been analysed quantitatively and found to be consistent with the concept of the membrane which has been outlined.

References

[1] Blake, J.R. and Sleigh, M.A. (1974), *Biol. Rev.*, **49**, 85–125.

[2] Brokaw, C.J. (1965), *J. exp. Biol.*, **43**, 155–169.

[3] Brokaw, C.J. (1971), *J. exp. Biol.*, **53**, 445–64.

[4] Evans, E.A. (1973), *Bioph. J.*, **13**, 926–940 and 941–954.

[5] Holwill, M.E.J. (1974), *Sci. Prog. Oxf.*, **61**, 63–80.

[6] Rappaport, R. (1967), *Science, N.Y.*, **156**, 1241–3.

[7] Rikmenspoel, R. (1971), *Biophys. J.*, **11**, 446–63.

[8] Summers, K.E. and Gibbons, I.R. (1971), *Proc. natn. Acad. Sci. U.S.A.*, **68**, 3092–6.

[9] Weis-Fogh, T. and Amos, W.B. (1972), *Nature, Lond.*, **236**, 301–304.

[10] Wessells, N.K. (1971), *Scient. Am.*, **225**(4), 76–82.

[11] Yoneda, M. and Dan, K. (1972), *J. exp. Biol.*, **57**, 575–88.

Conclusions

Most of this book is about things we know, or think we know. This last page points to things we don't know. Many gaps in our knowledge are evident from the preceding chapters. Most of our understanding of biological structural materials is based on analogy with man-made materials. Theories developed to explain the properties of plastics and of composite materials have been tremendously helpful to biologists but concepts originating within biomechanics, to explain the peculiar features of biological materials, are still few. We do not understand fully how vertebrate striated muscle works, and other important types of muscle are even less well understood. Our knowledge of the stresses which muscles develop in living animals, and the stresses which skeletons have to withstand, is limited to a few examples. The thinning of the boundary layer which apparently results from the tail movements of swimming fish has not been fully accounted for (p. 37). Our understanding of joint lubrication is still rather hazy and treatment of osteoarthritis seems still to be based more on trial and error than on physical theory (p. 39). The very existence of negative pressure in xylem has recently been questioned (p. 47).

There is another group of problems which has captured the interest of a group of zoologists led by Knut Schmidt-Nielsen. These are prob-lems of scaling, that is of the effects of absolute size on the structure and capabilities of animals [1], [3]. Such a problem was attacked by Galileo when he enquired why large animals (such as elephants) have relatively thicker bones than small ones (such as mice) [2]. His answer was that strength is proportional to cross-sectional area, and that unit cross-sectional area of bone would have to support more weight in large animals if it were not for the change in proportions. This is only a partial answer, since the legs of small animals can exert forces of many times the weight of the body. Another scaling problem is presented by Fig. 4.3 which shows that the cost of transport for quadrupedal mammals is rather precisely proportional to (body weight)$^{-0.4}$. So far, this has not been ex-plained.

References

[1] Alexander, R.McN. (1971), *Size and Shape*, Edward Arnold, London.
[2] Galileo Galilei. (1638); *Discorsi e dimonst-razioni matematiche intorno a due nuove scienze*, Elsevir, Leiden. Translated, 1954, as *Dialogues concerning two new sciences*. Dover, New York.
[3] Schmidt-Nielsen, K. (1970), *Fed. Proc.,* **29**, 1524–1532.

Index

Abductin, 9
Arteries, 42–43

Blood and blood vessels, 40–43
Bone, 12–14, 19, 20
Bones, tubular, 23
Branches, 22
Brett, J.R., 26

Capillaries, 41
Cell division, 56
Cell membranes, 58–59
Cilia, 53–56
Collagen, 11
Composite materials, 11–14
Conveyor belt, 25–26
Corpuscles, red, 40–41, 58
Cost of transport, 27–28
Cuticle, insect, 11

Dogs, jumping by, 17–19, 32–33
Dowson, D., 40

Echinoderm ossicles, 12
Elasticity, 9, 31–33
Elastin, 10
Exoskeleton, 23–24

Fibres, 10–11
Fish feeding, 20–21
Flagella, 53–56
Flea, 21
Flight, 21, 27, 33–36
Force platform, 17–18, 28–29

Gliding, 33–34

Hovering, 34–36

Joints, lubrication of, 39–40
Jumping, 17–19, 21, 32–33

Kangaroo, 27, 32

Lungs, 43–45

McCutchen, C.W., 40
Muscle, 14–15, 18–19, 20–21
Muscles, pennate, 18, 22–23
Myonemes, 57–58

Oxygen consumption, 25–27

Pigeon, flight of, 21
Pits, 48–49
Polymers, 8
Polysaccharides, 8
Pressure transducer, 20
Proteins, 8

Resilience, 10
Resilin, 9–10, 21
Rubbers, 9
Running, 26–27, 28–33

Sap, 46–49
Scaling, 60
Scallop, 9
Schmidt-Nielsen, K., 26, 60
Scholander, P.F., 47
Seeds, falling, 50–52
Silk, 11
Stomata, 49–50
Strain gauge, 20
Surfactants, 44–45
Swimming, 26–27, 36–37

Taylor, C.R., 26
Trees, 22, 46–49
Tucker, V.A., 26

Units, S.I., 7

Veins, 42
Viscosity, 8–9
Vorticella, 57

Walking, 27
Water tunnel, 26
Weis-Fogh, T., 9, 34

Wind tunnel, 26
Wood, 11

Young's modulus, 10

Zoothamnium, 57

Baby Pet Animals